T0235078

Algorithm-Driven Truss Topology Optimization for Additive Manufacturing

Christian Reintjes

Algorithm-Driven Truss Topology Optimization for Additive Manufacturing

Christian Reintjes
Dielheim, Germany

Supervisor and first appraiser
Prof. Dr. rer. nat. Ulf Lorenz
University of Siegen
Second appraiser
Prof. Dr.-Ing. Peter F. Pelz
Technical University of Darmstadt
Date of the oral examination
26. August 2021

ISBN 978-3-658-36210-2   ISBN 978-3-658-36211-9 (eBook)
https://doi.org/10.1007/978-3-658-36211-9

Responsoible Editor: Stefanie Eggert
This Springer Vieweg imprint is published by the registered company Springer Fachmedien
Wiesbaden GmbH part of Springer Nature.
The registered company address is: Abraham-Lincoln-Str. 46, 65189 Wiesbaden, Germany

Acknowledgments

The first and foremost thanks goes to my advisor Prof. Dr. rer. nat. Ulf Lorenz for his valuable mathematical and engineering input and continuous support during the development of this thesis. I am especially grateful for the freedom you gave me to build my research specialization in the field of algorithm-based optimization—without you, I probably would not have ended up in the field of applied mathematics. I will remember with great gratitude the interdisciplinary research between engineering sciences and mathematics, the possibilities of interpretation associated with additive manufacturing and component design, and the exciting and valuable discussions about it.

I also thank Prof. Dr.-Ing. Peter F. Pelz for acting as a referee for this thesis. Further, I thank Prof. Dr.-Ing. Bernd Engel and Prof. Dr. rer. nat. Robert Brandt for taking the time out to act as examiners in my examination committee.

I gratefully acknowledge the financial support from the University of Siegen through the SME Graduate School and the opportunity to start research parallel to my master's studies and be supervised by experienced researchers.

I am particularly grateful for the support and good times provided by my colleagues at the Chair of Technology Management. Special thanks goes to Alexander, Jonas, Tobias, and Michael.

Special thanks to my parents: I would like to take this opportunity to thank them wholeheartedly for their unconditional support. Last, but by no means least, I want to thank my wife, Janina. Words are powerless for expressing my gratitude.

Zusammenfassung

Unter dem Aspekt, dass die additive Fertigung (AF) die Herstellung geometrisch komplexer Gitterstrukturen realisiert – im Gegensatz zu dem Großteil der konventionellen Fertigungsverfahren – gewinnt die Kombination von Leichtbau, Topologieoptimierung und AF zunehmend an Bedeutung. Neben den etablierten Methoden der Topologieoptimierung wird der algorithmenbasierten Optimierung auf Basis der linearen Optimierung bislang wenig Aufmerksamkeit geschenkt. Um diese Forschungslücke zu schließen, wurden in der vorliegenden Arbeit die lineare Optimierung, das rechnergestützte Konstruieren (engl. CAD) und die numerische Formoptimierung und Simulation zu einer durchgehenden Prozesskette für die Entwicklung additiv gefertigter Gitterstrukturen kombiniert. Mit der entwickelten Ansys SpaceClaim Erweiterung construcTOR steht die hohe Leistungsfähigkeit von mathematischen Optimierungsalgorithmen, die in Optimierungslösern wie CPLEX implementiert sind, nun als Entwicklungswerkzeug in einer bestehenden 3D-CAD-Umgebung zur Verfügung. Die grafischen Möglichkeiten der 3D-CAD-Umgebung werden zur mathematischen Modellbildung genutzt. Aus den Optimierungsergebnissen werden für die Fertigung aufbereitete 3D-CAD Daten generiert.

In dieser Arbeit wurden drei *gemischt-ganzzahlige lineare Programme* (engl. MILPs) und ein robustes *quantifiziertes gemischt-ganzzahliges lineares Programm* (engl. QMIP) entwickelt. Diese Programme ermöglichen eine lineare Optimierung, die sich auf a) Konstruktionsregeln und Fertigungsrestriktionen für die AF, b) stützstrukturfreie Gitterstrukturen und c) Unsicherheit in der Belastung des zu optimierenden Bauteils konzentriert. Die ersten beiden MILPs und CPLEX wurden verwendet, um eine räumliche Gitterstruktur mit 1940 Strukturelementen mit und ohne Berücksichtigung von Stützstrukturen zu optimieren. Das dritte

MILP berücksichtigt fertigungsprozessbedingte minimale Querschnitte von Strukturelementen und Restriktionen um Ebenensymmetrie zu gewährtleisten. Das Programm wurde für die Optimierung eines additiv gefertigten Demonstrators eines Niederhalters in Leichtbauweise genutzt.

Weiterhin wurde ein robustes QMIP formuliert. Anstatt den kritischen Belastungsfall aus der möglichen Überlagerung aller Belastungsfälle zu ermitteln und diesen anschließend zu optimieren, konnten wir mit unserem robusten QMIP jegliche Kombination von Belastungsfällen, inklusive des zuvor nicht ermittelten kritischen Belastungsfalls, in der Optimierung berücksichtigen. In dieser Arbeit wurden Ergebnisse zu einer räumlichen Gitterstruktur mit 296 Strukturelementen unter Berücksichtigung von sieben Belastungsfällen und 128 Szenarien für die mathematische Optimierung präsentiert. Außerdem wurden die Ergebnisse zu einer räumlichen Gitterstruktur mit 1720 Strukturelementen mit acht einzeln auftretenden Belastungsfällen präsentiert. Anhand der Optimierung von additiv gefertigten Gitterstrukturen haben wir gezeigt, dass unser algorithmenbasierter Konstruktionsprozess ein effizientes und zuverlässiges Werkzeug für die Entwicklung von additiv gefertigten Gitterstrukturen ist.

Abstract

Since Additive Manufacturing (AM) techniques allow the manufacture of complex-shaped structures—compared to most conventional manufacturing techniques—the combination of lightweight construction, topology optimization, and AM is of significant interest. Besides the established continuum topology optimization methods, less attention is paid to algorithm-driven optimization based on linear optimization, which can also be used for topology optimization of truss-like structures.

To overcome this shortcoming, we combined linear optimization, Computer-Aided Design (CAD), numerical shape optimization, and numerical simulation into an algorithm-driven product design process for additively manufactured truss-like structures. With our Ansys SpaceClaim add-in construcTOR, which is capable of obtaining ready-for-machine-interpretation CAD data of truss-like structures out of raw mathematical optimization data, the high performance of (heuristic-based) optimization algorithms implemented in commercial linear programming software such as CPLEX is now available to the CAD community. Against this background, three Mixed-Integer Linear Programs (MILPs) and one robust Quantified Mixed Integer Linear Program (QMIP) are developed. Our models allow linear optimization focusing on a) design rules, limitations and standards for AM, b) support-free truss-like structures, and c) loading uncertainty.

Our first two MILPs and CPLEX were used to optimize a spatial truss-like structure consisting of 1940 structural members suitable for powder-based AM with and without enforcing a self-supporting truss-like structure. We further applied our algorithm-driven product design process to the real-world application of designing an additively manufactured lightweight forming tool. We used our third MILP and CPLEX considering minimal cross-sectional areas of the structural members depending on AM limitations and standards, and a design-variable linking technique to enforce two-fold symmetry.

In addition, we utilized a robust QMIP formulation. Instead of determining and optimizing a single worst-case scenario, our approach allowed us to state multiple loading cases while ensuring that the resulting structure is stable, even in the (unknown) worst-case scenario. We have presented results on a spatial truss-like structure consisting of 296 structural members by considering the combination of seven loading cases, which resulted in 128 loading scenarios and a spatial truss-like structure consisting of 1720 structural members with eight individually occurring loading cases. We demonstrated using real-world design cases to show that our algorithm-driven product design process is an efficient and reliable optimization tool for preliminary designs of truss-like structures.

Contents

Symbols[1]

A_b	L^2 / cross-sectional area of a structural member b
b	structural member
b_g	L / minimum gap size for immovable components
b_o	L / self-supporting overhang of a component
β	transition angle
D	L / outer diameter of a self-supporting structural member independent of the component orientation
D_h	L / outer diameter of a self-supporting structural member dependent of the horizontal component orientation
D_t	pre-processed dimension of a structural member b of type t
D_v	L / outer diameter of a self-supporting structural member dependent of the vertical component orientation
δ	downskin angle
δ_{cr}	critical downskin angle
E	set of edges
E_b	$ML^{-1}T^{-2}$ / Young's modulus
F_{ij}	LMT^{-2} / force vector between two directly adjacent nodes i, j
$F_W^G p_{x,y,z}$	global coordinate system of the reference volume $\mathbb{V}$
$F_W p_{x,y,z}$	local coordinate system of a structural member $B_{t,i,j}$
G_b	$ML^{-1}T^{-2}$ / shear modulus of a structural member b
γ	safety factor
I_b	L^4 / area moment of inertia of a structural member b

---

[1] The symbols in the first column are described in the second column. The second column of the table, if available, indicates the dimension as a monomial with the primary dimensions mass (M), length (L), and time (T).

k	set of different loading cases where the self-weight loads are considered
k_b	Timoshenko's shear coefficient
L_b	L / length of a structural member b
L_b^0	L / undeformed length of a structural member b
m_k	ML^2T^{-2} / one of two moments acting centrically at one of two adjacent nodes
m_l	ML^2T^{-2} / one of two moments acting centrically at one of two adjacent nodes
M_t	pre-processed material constitution of a structural member b of type t
$\mathbb{A}$	L^3 / assembly space
$\mathbb{D}$	L^2 / downskin area
$\mathbb{G}$	ground structure
$\mathbb{U}$	L^2 / upskin area
$\mathbb{V}$	L^3 / reference volume
μ_b	Poisson's ratio of a structural member b
N	number of layers needed for a structural member b of type t
v_b	L^3 / volume of a structural member b
B	compatibility matrix
g_b	LMT^{-2} / nodal gravitational force vector due to self-weight and external loads
p	LMT^{-2} / nodal force vector of the free degrees of freedom
q	LMT^{-2} / member force vector
ψ_b	central axis of a structural member b
Q_b	$ML^{-1}T^{-2}$ / stress constraint for both tension and compression
q_b^-	LMT^{-2} / member force compression
q_b^+	LMT^{-2} / member force tension
σ_b	$ML^{-1}T^{-2}$ / stress constraint for both tension and compression
T	truss
t_b	L^3 / structural member volume
τ	L / layer thickness
U	set of void structural members
$\bar{C}_{t,i,j}$	$ML^{-1}T^{-2}$ / upper bound of the stress of a structural member b of type t
υ	upskin angle
υ_{cr}	critical upskin angle
V	set of vertices
V_T	L^3 / volume of a truss T

x_{ij} binary variable indicating whether a structural member is present
            between nodes i, j
Z build direction
Z_b L^3 / plastic section modulus of a structural member b

Acronyms[2]

AM	Additive Manufacturing
API	Application Programming Interface
BREP	Boundary REPresentation
CAD	Computer-Aided Design
CAE	Computer-Aided Engineering
CAM	Computer-Aided Manufacturing
CP	Conic Programming
DEP	Deterministic Equivalent Problem
DfAM	Design for Additive Manufacturing
FDM	Fused Deposition Modeling
FEA	Finite Element Analysis
GUI	Graphical User Interface
ICP	Integer Conic Programming
ILP	Integer Linear Programming
IP	Integer Programming
ISDP	Integer SemiDefinite Programming
ISS	International Space Station
LP	Linear Programming
MILP	Mixed-Integer Linear Programming
MINLP	Mixed-Integer Nonlinear Programming
MIP	Mixed-Integer Programming
MIQP	Mixed-Integer Quadratic Programming
MISDP	Mixed-Integer SemiDefinite Programming
MISOCP	Mixed-Integer Second Order Cone Programming

---

[2] The acronyms in the first column are described in the second column.

NLP	Nonlinear Programming
OR	Operations Research
QMIP	Quantified Mixed-Integer Programming
RAM	Random Access Memory
RTTO	Robust Truss Topology Optimization
SDP	SemiDefinite Programming
SIMP	Solid Isotropic Material with Penalization
SLA	StereoLithogrAphy
SLM	Selective Laser Melting
SLS	Selective Laser Sintering
STEP	Sandard for The Exchange of Product model data
STL	Standard Triangle Language
TOR	Technical Operations Research
TTD	Truss Topology Design
TTO	Truss Topology Optimization
UI	User Interface

List of Figures

List of Tables

Introduction

1

Every kind of science, if it has only reached a certain degree of maturity, automatically becomes a part of mathematics.

(David Hilbert)

The ideal engineer is a composite... He is not a scientist, he is not a mathematician, he is not a sociologist or a writer; but he may use the knowledge and techniques of any or all of these disciplines in solving engineering problems.

(Nathan Washington Dougherty)

1.1 Motivation

We are living in an age of optimization. Most people strive for the most efficient and low-risk route to the destination calculated by their car navigation system, affordable housing with a holistic energy-efficient design, and a fast as well as energy-efficient car, e.g., with optimized passive safety in the cockpit area concerning crashworthiness. Solving large-scale real-world optimization problems is ubiquitous in engineering optimization, to design and manufacture products both efficiently and economically.

The beginnings of Linear Programming (LP) can be traced to the 1940s, when it was developed following research on military operations. With the advancement of computers and software technology in the 1950s and 1960s, Linear Programming (LP) and Mixed-Integer Linear Programming (MILP) evolved and found different applications in the field of Operations Research (OR), e.g., production scheduling, transportation, facility location, and strategic allocation of resources (see, e.g.,

Dantzig 1963, Dantzig et al. 1959, Eisemann 1955, Grigoriadis 1966, Orchard-Hays 1968). Nowadays, based on the high performance of efficient LP solvers, such as CPLEX (CPLEX 2019), large-scale (mixed-integer) LPs of practical relevance with up to millions of variables and constraints can be solved, which was inconceivable even a few decades ago. Thus, the application of OR techniques, employed successfully, e.g., in supply chain management, gain importance in engineering optimization, in addition to the existing solution techniques from nonlinear, geometric, dynamic, and stochastic programming (Rao 2019). In recent years, attempts have been made (see, e.g., Morsi et al. 2012, Pöttgen et al. 2016, Weber and Lorenz 2017) to adapt and implement these OR techniques to engineering optimization. The approach of applying OR techniques to technical (engineering) applications is called Technical Operations Research (TOR). The TOR approach has been widely applied to heterogeneous technical systems, e.g., ventilation systems (Schänzle et al. 2015), heating systems (Pöttgen et al. 2016), gearboxes (Dörig et al. 2016), booster stations (Weber and Lorenz 2017), thermofluid systems (Weber et al., 2020), pumping systems (Müller et al., 2020), truss structures (Gally et al., 2015, 2018, Kuttich, 2018, Mars, 2014), additively manufactured truss-like structures (Reintjes and Lorenz 2020), and wire-arc Additive Manufacturing (AM) (Bähr et al. 2020).

An emerging sub-domain in the research field of TOR is the application of (mixed-integer) LP techniques to Truss Topology Optimization (TTO), which focuses on engineering design practice and Design for Additive Manufacturing (DfAM) exploiting the opportunities provided by AM technologies. At the time of writing, TTO for AM is frequently managed using nonlinear optimization techniques like the established and in industrial finite element software implemented continuum topology optimization methods based on the Solid Isotropic Material with Penalization (SIMP) (see, e.g., Saadlaoui et al. 2017). Contrary to nonlinear optimization techniques, linear optimization techniques have the advantage of being able to ensure that optimization algorithms cannot get stuck in a non-global, i.e., local, optimum. This advantage far outweighs the disadvantage that the (mixed-integer) LPs cannot consider nonlinear constitutive material equations, e.g., linear elastic material behavior given by nonlinear constraints, without costly linearization. To maximize the advantage and compensate for the disadvantage, it is natural to reformulate nonlinear TTO programs into (mixed-integer) LPs. Our objective is to generate preliminary designs of truss-like structures applying mathematical optimization and afterwards use linear elastic and nonlinear elastic numerical analysis (Finite Element Analysis (FEA)) to verify and validate the mathematical optimization solutions. Consequently, the efficient (mixed-integer) LP solvers' high performance, i.e., inter alia the profound experience and broad knowledge of the research fields mathematical programming and OR about complexity theory and heuristics, becomes available for the research field TTO for AM. In combination with our Ansys SpaceClaim add-in `constructOR`, which allows generating algorithm-driven design studies within the Ansys workbench (ANSYS 2019), the (mixed-integer) LP solvers' high performance also becomes available to Computer-Aided Design (CAD) engineers. In this thesis, the application of TOR to TTO for AM is investigated. Our objective is to find a truss-like structure with minimal volume such that, for a anticipated external static, i.e., time-independent, loading scenario, the truss-like structure remains in a static equilibrium position. This formulation (task) is equivalent to a minimum weight truss design formulation with structural members (Gross et al. 2012), e.g., bars or beams, subjected to stress constraints. The simultaneous size, shape, and topology optimization of planar and spatial trusses is stated using three MILPs and one (robust) Quantified Mixed-Integer Programming (QMIP) that allow optimization focusing on a) design rules, manufacturing limitations and standardization for additively manufactured truss-like structures, b) support-free truss-like structures, and c) loading uncertainty, or any combination thereof.

To the best of the authors' knowledge, 3D-CAD software are not tailored for mathematical optimization based TTO since no functions to generate 3D-CAD models from mathematical optimization data, for instance the CPLEX solution pool, exists. To solve this issue and realize an algorithm-driven design workflow for AM, i.e., from mathematical optimization to the manufactured component, we developed the Ansys SpaceClaim add-in `constructOR` to generate 3D-CAD models of our optimized truss-like structures (mathematical optimization solutions) inclusive of geometry cleanup and simplification for FEA. The automatic 3D-CAD assembly construction approach considers engineering design practice and generates a 3D-CAD model out of raw mathematical optimization data ready-for-machine-interpretation. The capabilities and current limits of the proposed algorithm-driven design workflow for AM and the included mathematical optimization models are demonstrated in numerous design studies.

1.2   Structure of the Thesis

The rest of this thesis consists of seven chapters, which are organized as follows:

- **Chapter 2: Physical and Technical Background.** This chapter covers the physical and technical foundations of TTO. The fundamentals of structural mechanics focusing on the linear theory of structural members, AM technologies, and support structures for AM are presented. The chapter concludes with an overview of mathematical programming focusing on MILP.

- **Chapter 3: Optimization of Truss Structures.** The general form of a TTO problem and existing solution approaches are reviewed. Furthermore, the ground structure approach is introduced. In the following, complexity in structural optimization, in the sense of intellectually challenging designs of advanced structures, does not necessarily coincide with computational complexity, but interdependencies are elaborated. The same is done for symmetry in structural optimization and in mathematical design optimization. Furthermore, a two-tier overview of related work, seen from the perspective of mathematical optimization and structural optimization, is presented. Why TTO is particularly suitable for combining structural optimization with mathematical optimization and how it can profit from mathematical optimization techniques are discussed.

- **Chapter 4: Bridging Algorithm-Driven Truss Optimization and Additive Manufacturing.** Details on how the TOR approach was adapted to develop an algorithm-driven product design process for additively manufactured truss-like structures are presented. As a main contribution of this chapter, we accomplish a pre-processing technique that enables us to define a realistic upper bound of the structural member's stress under consideration of a factor of safety. Furthermore, how the build volume of an AM system is represented in relation to the machine origin using the ground structure approach is shown.

- **Chapter 5: Mixed-Integer Linear Programming for Truss Optimization.** This chapter introduces three MILP and one (robust) QMIP minimum weight truss design formulations with structural member stress constraints. To move closer to actual AM application, the design rules, manufacturing limitations, AM technology standards, and support structures for AM are modeled using DfAM relevant linear constraints.

- **Chapter 6: Design and Implementation of a CAD Tool for Mathematical Programming.** Implementation details of our data processing framework, the CAD neutral Ansys SpaceClaim add-in `construcTOR` inclusive of the Graphical User Interface (GUI), and post-processing algorithms are presented. In a detailed computational study, the performance of modeling multiple structural members depending on the geometrical complexity of the structural member's cross-section are evaluated. In addition, post-processing of interferences from clashing bodies and the merging process of all bodies, based on raw mathematical optimization data, are evaluated.

- **Chapter 7: Computational Study.** A series of design studies are extensively investigated to demonstrate the performance and real-world applicability of the proposed algorithm-driven product design process for additively manufactured truss-like structures.

- **Chapter 8: Conclusion and Outlook.** A summary of the main research findings and implications for future research in the field of algorithm-driven TTO for AM conclude the thesis.

1.3   Own Contribution to Knowledge

The main contribution of this thesis can be summarized as follows:

- We give a broad introduction to the topic TTO for AM and highlight discrepancies between mathematical optimization and real-world engineering application. We show that small changes in the TTO optimization models, which are necessary and ordinary from an engineering perspective, in general, are a more difficult task in mathematical optimization and can come at the expense of an increased computational complexity. We further shed light on the algorithm-driven TTO for AM and encourage actual engineering application by supporting structural optimization to adopt the long-standing experience in theory, heuristics, and modeling of mathematical optimization.

- We present an algorithm-driven truss optimization workflow to fully exploit the advantages of mathematical optimization combined with AM. Mathematical optimization, finite element shape optimization, and FEA are used within the optimization workflow. Mathematical optimization is used to generate preliminary designs of truss structures. Finite element shape optimization is used to minimize the stress response of high-stressed connection nodes. FEA is only used for validation.

- We develop three MILP and one (robust) QMIP minimum weight truss design models with structural member stress constraints. All models are implemented with and without single or double axial symmetry. Each model focuses on specific technical challenges of DfAM. The MILP $TTO_{1;p}$ focuses on powder-based DfAM, MILP $TTO_{1;s}$ on support-free truss-like structures, and the MILP $TTO_{1;m}$ on design for manufacturability. The QMIP $TTO_{1;q}$ extends our minimum weight truss design models to loading uncertainty. Hence, the worst-case structural analysis becomes less important.

- We design and implement the Ansys SpaceClaim add-in `construcTOR`, a for the most part automated software for mathematical design optimization. We enable CAD engineers to generate algorithm-driven design studies within the CAD neutral Ansys Workbench without profound knowledge about mathematical optimization. The add-in involves a GUI and bidirectional connection to CPLEX and all major 3D-CAD systems. It transforms CAD neutral design studies into large-scale mathematical optimization instances, and the optimization results into a 3D-CAD model. We develop an efficient algorithm to post-process the intersection of structural members and an algorithm to merge all structural members. We show that the implementation of our algorithm can massively speed up the merging process.

- We provide several algorithm-driven design studies for all three MILPs and the QMIP and partially verify and validate the mathematical optimization solutions via FEA and additively manufactured functional prototypes. As a real-world design study representing the whole algorithm-driven design workflow for AM, the optimization of an additively manufactured lightweight forming tool is presented inclusive of subsequent finite element shape optimization of high-stressed connection nodes of the optimized truss-like structure.
- This thesis is supplemented by the software construcTOR. The sources of the software and the data that support the findings of this study are available from the corresponding author, C. Reintjes, upon reasonable request.

Partial results of this thesis have been published in the following set of publications. The work was completed in close cooperation with the respective co-authors.

As lead author:

- C. Reintjes, M. Hartisch and U. Lorenz. Lattice structure design with linear optimization for additive manufacturing as an initial design in the field of generative design. In *Operations Research Proceedings 2017: Selected Papers of the Annual International Conference of the German Operations Research Society (GOR)*, pages 451–457, 2018.
- C. Reintjes, M. Hartisch and U. Lorenz. Design and optimization for additive manufacturing of cellular structures using linear optimization. In *Operations Research Proceedings 2018: Selected Papers of the Annual International Conference of the German Operations Research Society (GOR)*, pages 371–377, 2019.
- C. Reintjes and U. Lorenz. Mixed integer optimization for truss topology design problems as a design tool for AM components. In *International Conference on Simulation for Additive Manufacturing*, 2:193–204, 2019.
- C. Reintjes, M. Hartisch and U. Lorenz. Support-free lattice structures for extrusion-based additive manufacturing processes via mixed-integer programming. In *Operations Research Proceedings 2019: Selected Papers of the Annual*

International Conference of the German Operations Research Society (GOR), pages 465–471, 2020.

- C. Reintjes and U. Lorenz. Bridging mixed integer linear programming for truss topology optimization and additive manufacturing. In *Optimization and Engineering*, pages 1–45, 2020.
- C. Reintjes and U. Lorenz. Algorithm-driven optimization of support-free truss-like structures in early-stage design for additive manufacturing. In *PAMM: Proceedings in Applied Mathematics and Mechanics*, 20: e202000204, 2021.
- C. Reintjes, J. Reuter, M. Hartisch, U. Lorenz and B. Engel. Towards CAD-based mathematical optimization for additive manufacturing—designing forming tools for tool-bound bending. In *Pelz P.F., Groche P. (eds) Uncertainty in Mechanical Engineering. ICUME 2021. Lecture Notes in Mechanical Engineering*, pages 12–22, 2021

As co-author:

- M. Hartisch, C. Reintjes, T. Marx and U. Lorenz. Robust topology optimization of truss-like space structures. In *Pelz P.F., Groche P. (eds) Uncertainty in Mechanical Engineering. ICUME 2021. Lecture Notes in Mechanical Engineering*, pages 296–306, 2021

Physical and Technical Background

2

This chapter provides the physical and technical background for algorithm-driven truss-like[1] topology optimization for AM. We start in Section 2.1 with the fundamentals of structural mechanics, focusing on the one-dimensional linear theory of structural members and trusses in general. Afterwards, in Section 2.2, we introduce AM technologies and support structures. We conclude this chapter with a brief overview of mathematical programming in Section 2.3, introducing graph theory, MILP, and the optimality gap.

2.1 Structural Mechanics

Mechanics is the science of force and motion of matter, wherein solid mechanics is the science of force and motion of matter in the solid state (Fung et al. 2017). As a central part of continuum mechanics, solid and structural mechanics provides the theoretical basis for the linear theory of structural members (Hjelmstad 2005). In this thesis, we mainly concentrate on the material models for three-dimensional linearized elasticity problems (see, e.g., Slaughter 2012), vital for the linear theory of structural members. The derived principles most relevant for this work are briefly explained below, with the explanations presented based on the works of

---

[1]We use the term *truss-like structures* as a synonym for truss structures in the context of DfAM and mathematical programming, rather than solely referring to trusses as it is commonly used in structural engineering.

Supplementary Information The online version contains supplementary material available at https://doi.org/10.1007/978-3-658-36211-9_2.

Coates et al. (2019), Gross et al. (2012), Hjelmstad (2005), Mang and Hofstetter (2013). As for prerequisites, the reader is expected to be familiar with scalars, vectors, tensors, and strain-displacement relations.

2.1.1   Three-Dimensional Continuum Mechanics

Newton's law of conservation of linear and angular momentum leads to the *equilibrium of forces and moments* (see, e.g., Gross et al. 2012) in a static (time-independent) context, considering *continuous bodies*, i.e., bodies modeled as a continuous system and not as a discrete particle system. In order to introduce a concept to model the force transmission through a deformable continuous body, let us introduce the notion of *stress* and *traction*, as given by Cauchy (see, e.g., Irgens 2008).

The Traction Vector and the Stress Tensor
Hjelmstad (2005) emphasizes that the connection between traction forces and stresses lies in the fact that the state of stress of an *isolated free-body* must be represented as equivalent traction to claim equilibrium of forces and moments. Consider a wafer with thickness ϵ, diameter h, lateral contour Γ, and surface Ω, whereby $h \gg \epsilon$ holds; see Figure 2.1. Let t_n, t_Γ, b be the signed traction field (force per unit area), the average traction through the thickness ϵ of the wafer with the lateral contour being Γ, and a body force acting on the volume, respectively (Hjelmstad 2005). We will use the symbol n to denote the signed normal vector for the plane on which a traction force is acting. Thus, $t_n \, dA$ (force per unit area times area) is the total force acting on the exposed surface, with dA being an (infinitesimal) area of this exposed surface. Since

$$h \gg \epsilon \tag{2.1}$$

holds, we can average the body force b over the thickness ϵ so that the body force $\epsilon \, b$ per unit area is acting on the wafer. The lateral contour Γ has tractions $\epsilon \, t_\Gamma$ per unit length (Hjelmstad 2005). We adhere to the convention that the *static equilibrium*

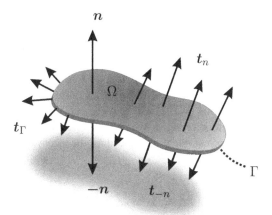

Figure 2.1 Isolated free-body of a wafer visualizing Cauchy's reciprocal theorem. (Obtained from Hjelmstad 2005)

$$\int_\Gamma \epsilon\, t_\Gamma\, ds + \int_\Omega t_n\, dA + \int_\Omega t_{-n}\, dA + \int_\Omega \epsilon b\, dA = 0 \qquad (2.2)$$

applies, where ds is the infinitesimal of the curve of the lateral contour and dA the infinitesimal of the surface of the body. For $\epsilon \to 0$, the limit being assumed to exist for Equation (2.2), such that

$$\int_\Omega (t_n + t_{-n})\, dA = 0 \qquad (2.3)$$

applies. We see at once that negligible tractions along the contour $\epsilon\, t_\Gamma$, first integral in Equation (2.2), and a negligible body force acting on the wafer $\epsilon\, b$, second integral in Equation (2.2), arise. Since the wafer's surface Ω can be chosen arbitrarily, the integral form simplifies to

$$t_n = -t_{-n}\,. \qquad (2.4)$$

In this way, we obtain what is known as *Cauchy's reciprocal theorem*, which is nothing but the statement that the traction on the area with normal n is the negative of the traction on the area with normal $-n$. This obvious intuitive meaning is fundamental for one-dimensional mechanics, e.g., establishing a truss in equilibrium (Hjelmstad 2005).

To simplify notation, we continue to abbreviate vector multiplication and vector differentiation. We write $u \cdot v$, $u \times v$, $u \otimes v$, $div\, v$ for the dot product, the cross product, the tensor product of two vectors u and v, and the divergence of vector v, respectively (Hjelmstad 2005, pp. 3–12, p. 40). Consider the so-called *Cauchy tetrahedron* (Hjelmstad 2005, p. 105) with vertices a, b, c, lengths ϵ_1, ϵ_2, ϵ_3, and areas a_i with normal vectors $-e_i$, as shown in Figure 2.2. From the Cauchy tetrahedron and *Cauchy's reciprocal relations*, we deduce the eigenvalue decomposition, i.e., the *Cauchy stress formula*

$$t_n = \sum_{i=1}^{3} (n \cdot e_i) t_{e_i} \ . \tag{2.5}$$

Explicitly, this means that the traction t_n on the area with normal n is defined by three base tractions $t_{e_1}, t_{e_2}, t_{e_3}$, and that a relation between the traction on the faces and the tractions on the coordinate faces exists so as to maintain equilibrium. Using the tensor product, we may identify

$$(n \cdot e_i) t_{e_i} = (t_{e_i} \otimes e_i)\, n \tag{2.6}$$

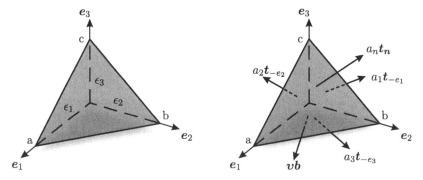

Figure 2.2 The Cauchy tetrahedron. (Obtained from Hjelmstad 2005)

and formulate the *stress tensor*[2]

$$S \equiv \sum_{i=1}^{3} t_{e_i} \otimes e_i \,, \tag{2.7}$$

which is crucially important in defining the transmission of forces through a continuous solid body. The principal significance of the stress tensor is that Equation (2.5) is nothing but the equation

$$t_n = Sn \,. \tag{2.8}$$

The preceding observation enables one to relate applied tractions on the surface of the body to the stress field inside the body. The crucial fact is that any stress state which satisfies these equations is an equilibrium stress state (Hjelmstad 2005).

Force Equilibrium
In order to state the static equilibrium of a continuous solid body, we set up notation and terminology of a *deformation map* $\phi(z)$ (Hjelmstad 2005, p. 64) characterizing the *deformed configuration* $\phi(\mathscr{B}_1)$ of an unstressed and unstrained body $\mathscr{B}_1$. Figure 2.3 details the deformation map. Using

$$x = \varphi(z) \tag{2.9}$$

with the position vector z relative to the coordinate system $\{x_1, x_2, x_3\}$, we can, in general, locate the position of a point in the deformed configuration $\phi(\mathscr{B}_1)$. This provides an intrinsic characterization of the deformation of a body. Given the exemplary (material) point $\mathscr{P}$ and curve $\mathscr{C}$ on body $\mathscr{B}_1$ with surface boundary Ω_1, we can define the same point $\phi(\mathscr{P})$ in the deformed configuration $\phi(\mathscr{B}_1)$. Furthermore, we will consider a second continuous solid body $\mathscr{B}_2$ with surface boundary Ω_2 constrained by a normal vector field n, traction vector field t_n, and body force field b, as depicted in Figure 2.4. Given the static equilibrium of the body $\mathscr{B}_2$, i.e., all forces acting on the body $\mathscr{B}_2$ must be equal to zero, it follows that

$$\int_{\Omega_2} t_n \, dA + \int_{\mathscr{B}_2} b \, dV = 0 \,. \tag{2.10}$$

---

[2] Note that S does not denote the second (symmetric) Piola-Kirchhoff stress tensor (see, e.g., Bonet and Wood 1997) as widely used in engineering.

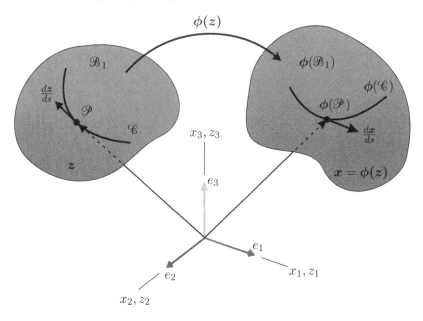

Figure 2.3 Characterization of the deformation of a continuous solid body: (left) reference configuration of the unstressed and unstrained body $\mathscr{B}_1$; (right) deformed configuration $\phi(\mathscr{B}_1)$. (Obtained from Hjelmstad 2005)

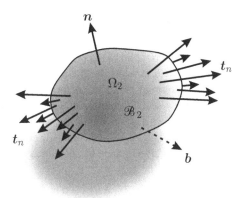

Figure 2.4 Continuous solid body $\mathscr{B}_2$ with surface boundary Ω_2 constrained to a normal vector field $\boldsymbol{n}$, traction field $\boldsymbol{t}_n$, and body force field $\boldsymbol{b}$. (Obtained from Hjelmstad 2005)

For simplicity of notation, we write dA, dV instead of the surface and volume integrals for a specific coordinate system (Hjelmstad 2005, p. 35). Applying Cauchy's stress formula (see Equations (2.5)–(2.8)) and Gauss's divergence theorem (see, e.g., Schey 2005), we get

$$\int_{\Omega_2} \boldsymbol{Sn}\, dA = \int_{\mathscr{B}_2} di\,v\boldsymbol{S}\, dV\,. \tag{2.11}$$

We can write Equation (2.10) in the form

$$\int_{\mathscr{B}_2} (di\,v\boldsymbol{S} + \boldsymbol{b})\, dV = \boldsymbol{0}\,. \tag{2.12}$$

Based on the concept of the *free-body diagram* (see Kövecses 2008a, b) and the fact that the equilibrium holds for any continuous solid body, Equation (2.12) becomes the local form of the equilibrium defining the rate of change of the symmetric stress tensor $\boldsymbol{S}$ as

$$di\,v\boldsymbol{S} + \boldsymbol{b} = \boldsymbol{0}\,. \tag{2.13}$$

Based on the preceding observation, the equilibrium stress state must be such that the traction vectors on the surface of the body $\mathscr{B}_2$ are equal to the applied traction vectors where they are prescribed (Hjelmstad 2005).

Moment Equilibrium
Our next concern will be the *balance of angular momentum*, defined by the requirement that the moment of the surface tractions and the body forces vanish. Let $\mathscr{B}_3$, Ω_3, $\boldsymbol{r}$ be a continuous solid body, the surface boundary of the body, and a position vector arranged from the origin of the coordinate system to the point with position vector $\boldsymbol{x}$, respectively. The balance of angular momentum is given by

$$\int_{\Omega_3} \boldsymbol{r} \times \boldsymbol{t}_n\, dA + \int_{\mathscr{B}_3} \boldsymbol{r} \times \boldsymbol{b}\, dV = \boldsymbol{0}\,. \tag{2.14}$$

Let $\boldsymbol{h}$ be an arbitrary, constant vector field. Under consideration of Equation (2.8) we can now proceed analogously to the proof and argumentation of (Hjelmstad 2005, pp. 113–114) and define a reduced form of the balance of angular momentum

$$\int_{\mathscr{B}_3} h \cdot [e_j \times Se_j] \, dV = 0, \qquad (2.15)$$

with an implied sum on j, where Se_j is subject to the condition

$$S_{ij} \equiv e_i \cdot Se_j = e_i \cdot \sum_{k=1}^{3} [t_{e_k} \otimes e_k] e_j = e_i \cdot t_{e_j}. \qquad (2.16)$$

In fact, we interpret the components of the stress tensor S, Equation (2.16), such that S_{ij} is the i-th component of the traction vector acting on the face with normal vector e_j. Assuming that the constant vector field h and region $\mathscr{B}_3$ can be chosen arbitrarily, the integral form of Equation (2.15) simplifies to

$$e_j \times Se_j = 0. \qquad (2.17)$$

Following Hjelmstad (2005), the above expression may be written as explicit relations

$$[S_{23} - S_{32}]e_1 + [S_{31} - S_{13}]e_2 + [S_{12} - S_{21}]e_3 = 0, \qquad (2.18)$$

which shows that since the base vectors are independent and nonzero, the only way the balance of angular momentum holds is if the components of the stress satisfy

$$S_{12} = S_{21}, \ S_{31} = S_{13}, \ S_{23} = S_{32}. \qquad (2.19)$$

Hence, in addition to Equation (2.14) being the balance of angular momentum, we require the stress tensor S to be symmetric, such that

$$S^T = S. \qquad (2.20)$$

Besides finding that we only need six independent quantities to define the state of stress at a (material) point of the continuous solid body rather than nine, we infer that the balance of angular momentum holds for any piece of the body, which is guaranteed by the symmetric stress tensor S.

2.1.2   One-Dimensional Linear Theory of Structural Members

Galileo Galilei (1564–1642) made the first contributions to structural member (beam) theory, mainly limited to the static equilibrium of structural members. Subsequently the mathematician and physicist Jakob Bernoulli (1654–1705) developed the *plane-sections* hypothesis, which asserts that plane sections remain plane and perpendicular to the midplane of a structural member after deformation and do not distort in their planes. In the spirit of Jakob Bernoulli, the mathematician Leonhard Euler (1707–1783) made vital advances to the plane-sections hypothesis. No less meaningful are the discoveries of Leonhard Euler relating the *theory of deflection* of curves of structural members, although no substantial improvements on *Bernoulli's kinematic hypothesis* were made. Claude Louis Marie Henri Navier (1785–1836) put the kinematic hypothesis right and thus laid the foundation for the structural member theory, one of the general theories of structural mechanics and structural analysis and therefore indispensable for structural engineers (Hjelmstad 2005).

According to Hjelmstad (2005), the concepts to describe the mechanics of a three-dimensional continuum body are hard to solve for a simple constitutive model like isotropic hyperelasticity (see, e.g., Weiss et al. 1996), even in the age of computers and FEA. A reduction from three-dimensional solid mechanics to two- or one-dimensional solid mechanics allows an equivalent reduction of the governing differential equations, reducing computational complexity (Arora and Barak 2009, Hjelmstad 2005). Reduced mechanical theories can characterize structural members in a one-dimensional manner as well as flat plates and curved shells in a two-dimensional manner.

The assumption on the linear theory of structural members is that the *linear elasticity theory* applies and, consequently, geometrical linearity and linear material behavior (Mang and Hofstetter 2013). Geometrical linearity requires small displacements, unlike the cross-sectional dimensions. Linear material behavior requires the validity of *Hooke's law*.

Prismatic Structural Member
Gordon (1978) defined a *structure* in solid mechanics as any assemblage of materials intended to sustain loads. An intuitive example inspired by nature is the topology of the human bone structure, which sustains the body load in an optimum way (Sigmund 1994), while it is still not widely understood how the microstructure of the bone adapts to mechanical stimuli and sustains loads (Huiskes and Hollister 1993). In compliance with Bauchau and Craig (2009), a structural member is a structure having one of its dimensions, length l, much larger than the other two cross-sectional dimensions. We assume that the cross-section is constant, so that the normal n_Γ to

the boundary of the cross-section Γ, see Figure 2.5, has no component in the axial direction x of the structural member.

In accordance with the notation shown in Figure 2.5, we define x, A, l to be the axis that coincides with the structural member's central axis defined by the structural member's local coordinate system, the cross-section, as well as the length of the structural member. The connecting line of all centers of gravity of the constant cross-sections coincides with the axis of the straight structural member (Mang and Hofstetter 2013).

Without loss of generality of the three-dimensional continuum mechanics, see Subsection 2.1.1, we assume that the motion or the tractions can be prescribed at the structural member ends ($x = 0$, $x = l$), with only the tractions on the lateral surface (Hjelmstad 2005).

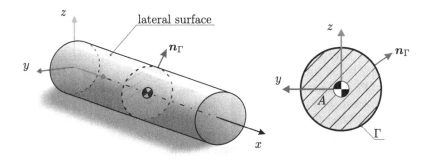

Figure 2.5 The terminology of a three-dimensional prismatic structural member. (Obtained from Hjelmstad 2005)

Axially Loaded Prismatic Structural Member

Let us denote by E, ν, N *Young's modulus* (Euler 1960), *Poisson's ratio* (Roylance 2008), and two equal in value *normal forces* of a prismatic structural member, respectively. Using *Saint-Venant's principle* (de Saint-Venant 1855, Toupin 1965) and requiring that no moment applies around the coordinate axes x lying in the cross-sectional plane of the structural member, the *principal axial stress* σ_x can be assumed to be uniformly distributed over the cross-section of the prismatic structural member, which implies

$$\sigma_x = \frac{N}{A}, \quad \text{where } \sigma_x \text{ is constant}. \tag{2.21}$$

Figure 2.6 details the state of stress for the special case of an axially loaded prismatic structural member using *Mohr's circle* (see, e.g., Parry 2004) with central point C. In the notation of Parry (2004) the maximum $\sigma_1 \equiv \sigma_x$ and the minimum $\sigma_2 = 0$ axial stress, the radius of the circle corresponds to the *minimal and maximal shear stress*

$$\tau_{min} = -\frac{N}{2A}, \quad \tau_{max} = \frac{N}{2A}. \tag{2.22}$$

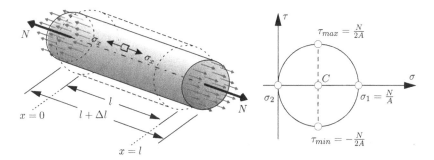

Figure 2.6 An axially loaded prismatic structural member with uniform axial stress σ_x (left) and associated state of stress represented by Mohr's circle (right). (Obtained from Parry 2004)

The *Euler-Bernoulli hypothesis* (see, e.g., Bauchau and Craig 2009), and assuming a prismatic structural member, leads to the *elongation* depending on the structural member's local coordinate system x, y, z

$$\varepsilon_x = \frac{\sigma_x}{E} = \frac{N}{EA}, \tag{2.23}$$

$$\varepsilon_y = \varepsilon_z = -\nu \frac{N}{EA}, \tag{2.24}$$

where ε_x, ε_y, ε_z are constant. The term EA defines the *axial rigidity* of a structural member. We are thus led to the structural member *extension* (see Figure 2.6 left)

$$\Delta l = \varepsilon_x l = \frac{Nl}{EA}. \tag{2.25}$$

Let us introduce the temporary notation M, $C \in \mathbb{N}$ for the set of all structural members of a truss-like structure and the set of any combination of multiple structural members with an identical central axis in a truss-like structure. See as an example Figure 2.9, the combination of structural members CG and GL. If we think of A^c, N^c, E^c, v^c $\forall c \in M$ as being constant for every single section of a structural member, we are in a position to write

$$\sigma_x^c = \frac{N^c}{A^c} \qquad\qquad \forall c \in C, \qquad (2.26a)$$

$$\tau_{min}^c = -\frac{N^c}{2A^c} \qquad\qquad \forall c \in C, \qquad (2.26b)$$

$$\tau_{max}^c = \frac{N^c}{2A^c} \qquad\qquad \forall c \in C, \qquad (2.26c)$$

$$\varepsilon_x^c = \frac{\sigma_x^c}{E^c} = \frac{N^c}{E^c A^c} \qquad\qquad \forall c \in C, \qquad (2.26d)$$

$$\varepsilon_y^c = \varepsilon_z^c = -v^c \frac{N^c}{E^c A^c} \qquad\qquad \forall c \in C, \qquad (2.26e)$$

$$\Delta l^c = \sum_{i \in c} \Delta l_i = \sum_{i \in c} \varepsilon_{x_i} l_i = \sum_{i \in c} \frac{N_i l_i}{E_i A_i} \qquad \forall c \in C. \qquad (2.26f)$$

A typical application for a structural design consisting of purely axially loaded prismatic members (bars or rods or both) is a truss; see Figure 2.7.

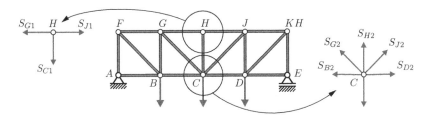

Figure 2.7 Plane static system of purely axially loaded structural members in the form of a truss including two free-body diagrams of equilibrium positions. (Obtained from Gross et al. 2012)

2.1.3 Truss Structures

In engineering, especially structural mechanics, trusses are among the most commonly used structures (Yan and Yam 2002). They are vitally important in multiple established industry areas, e.g., pylons, building exoskeletons, roof supports, towers, or in constructing bridges (Tejani et al. 2018a). Besides the listed industry areas, research concentrates on additively manufactured truss-like (lattice) structures for product design due to AM's capability to fabricate these complex microstructures (Johnston et al. 2006), which is not possible for traditional manufacturing. As additively manufactured connection nodes only consist of an accumulation of cured material, we assume, unless otherwise stated, that the connection node or joint is sufficiently dimensioned. In the next subsection we concentrate on *spatial trusses*, wherein, for better comprehensibility, we partially visualize *plane trusses*.

Classification of Trusses

A spatial truss (see Figure 2.8) consists of a quantity of structural members; each structural member is joined together at its ends (joints[3]) to the foundation, e.g., bearings, or to other structural members or to both, to form a stable (or rigid) three-dimensional structure. Frictionless joints are used (Coates et al. 2019).

Definition 2.1 (Simple Spatial Truss).
A Simple Spatial Truss exists at the initial state of six structural members and four joints, building a tetrahedron. Each successive extension consists of three new structural members and one joint.

Definition 2.2 (Compound Spatial Truss).
A Compound Spatial Truss consists of multiple simple spatial trusses joined together by six suitably arranged structural members to form a more complex truss.

Definition 2.3 (Complex Spatial Truss).
A Complex Spatial Truss is a truss that cannot be classified either as a simple or a compound truss.

Figure 2.8 Classification of trusses: (left) simple spatial truss; (middle) compound spatial truss; (right) complex spatial truss. (Obtained from Coates et al. 2019)

---

[3] In later chapters, we use the terms *(connection) node* or *vertex* or both as synonyms for joint as it is commonly used in mathematical optimization.

Assumptions of an Ideal Truss

Let an *ideal truss* (Gross et al. 2012) satisfy the following assumptions:

Assumption 2.1 (Connections).
The structural members are connected through frictionless joints.

Assumption 2.2 (Loadings and Reactions).
All loadings and reactions *are applied centrally at the joints.*

Assumption 2.3 (Centroids).
The centroid *for each structural member is straight and concurrent at a joint.*

Assumption 2.4 (External forces).
External forces *are either neglected, or their resultants are replaced by statically equivalent forces at the adjacent joints.*

Under the above assumptions, structural members of an ideal truss are subjected only to tension and compression and are therefore two-force structural members.

Zero-Force Structural Members

Zero-force structural members are those in a truss that only absorb forces during elastic deformation while maintaining a stable and internally determinate ideal truss. They are intended to decrease the risk of buckling (see Figure 2.11) by modifying the structural member's unsupported (buckling) length. See Figure 2.9, structural member CL divided into GL, GC by adding five[4] new structural members. The task of determining zero-force structural members can be solved by imposing the following observations (Coates et al. 2019, p. 93):

Observation 2.1.
If all except one of the structural members meeting at a joint lie in one plane, then the force in this odd structural member is zero if no external force acts at the joint. If an external force acts at the joint, then the force in the odd structural member must be such that its component normal to the plane of the other structural members is equal to the corresponding component of the external force (Coates et al. 2019).

Observation 2.2.
If three non-coplanar structural members meet at a joint not acted on by an external force, then the force in each structural member must be zero (Coates et al. 2019).

---

[4] As shown in Figure 2.9 (left) by the dashed lines GF, GH, GJ, GK, GM.

Observation 2.3.
If two of four non-coplanar structural members at a joint are collinear, then the forces in each of the two non-collinear structural members can be determined easily by considering force components normal to the plane containing the other three structural members (Coates et al. 2019).

Observation 2.4.
If all except two non-collinear structural members at a joint have zero force, then these two structural members must also have zero force if no external force acts at the joint (Coates et al. 2019).

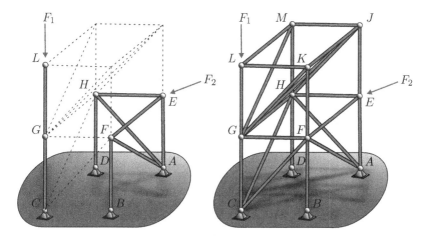

Figure 2.9 (Left) Ideal spatial truss acted on by a vertical force F_1 at joint L and force F_2 at joint E, including four bearings and dashed zero-force structural members; (Right) Same ideal spatial truss including zero-force structural members identified using Observations 2.1–2.4. (Obtained from Coates et al. 2019)

Determinacy

Let m, j, r temporarily stand for the number of structural members including the structural members mounted at the bearings, the number of joints j excluding the joints mounted at the bearings, and the number of unknown support reactions r, respectively. In line with Coates et al. (2019), an ideal truss is *externally statically determinate* if all the support reactions r can be determined using the conditions of static equilibrium; an ideal truss is *externally statically indeterminate* if the conditions of the static equilibrium are not sufficient to determine the support reactions r.

An ideal truss is *internally statically determinate* if the amount of structural members is sufficient to determine the forces in all structural members using Equations (2.4), (2.8), and (2.20); a truss is *internally statically indeterminate* if statics is not sufficient for determining the forces in all structural members. A truss is said to be *statically determinate* if the truss is both internally and externally statically determinate (Coates et al. 2019). A statically determinate truss is a *stable truss* given by a quantity of suitably mounted structural members, see Assumptions 2.1–2.4. These provide the natural and intrinsic characterization of a stable truss of which the shape cannot be modified without changing a structural member's length. Additional bearings or structural members are not necessary for determining the equilibrium configuration, which consequently yields a *degree of indeterminacy*. More precisely, an additional bearing leads to an *external redundancy*. An additional structural member leads to *internal redundancy*. Depending on the dimension $k \in \{2, 3\}$ of a truss, let $D_2 = 3$ and $D_3 = 6$ be the minimal amounts of support reactions needed to ensure restraint of a plane ($k = 2$) or spatial ($k = 3$) truss, with the property that

$$r \begin{cases} = D_k, & \text{externally determinate if truss is stable,} \\ > D_k, & \text{externally indeterminate truss,} \\ < D_k, & \text{unstable truss,} \end{cases} \qquad (2.27)$$

for the external determinacy and

$$3j \begin{cases} = m + r, & \text{internally determinate truss,} \\ > m + r, & \text{internally indeterminate truss,} \\ < m + r, & \text{unstable truss,} \end{cases} \qquad (2.28)$$

for the internal determinacy. The question naturally arises whether Equation (2.28) is a sufficient condition to design an ideal truss. The two small plane trusses ($r = 3$, $m = 9$, $j = 4$), see Figure 2.10 (Gross et al. 2012), demonstrate that Equation (2.28) is a necessary but not sufficient condition for stability, since we could find a truss layout that satisfies Equation (2.28), while simultaneously being improperly arranged. An improperly constrained truss may rotate about an angle ϕ and is statically indeterminate (Gross et al. 2012).

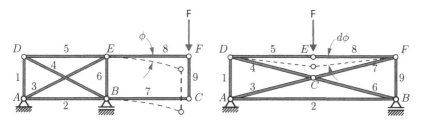

Figure 2.10 Improperly constrained trusses, according to (Gross et al. 2012)

The following assumptions for statically determinate (2.5 and 2.6) and statically indeterminate (2.7 and 2.8) trusses will be needed in the course of this thesis:

Assumption 2.5 (Statically Determinate Truss I).
Internal forces *are* not affected *by cross-sectional dimensions of structural members and Young's modulus.*

Assumption 2.6 (Statically Determinate Truss II).
Internal forces *are* not affected *by a slight lack of fit of a structural member or support displacement.*

Assumption 2.7 (Statically Indeterminate Truss I).
Internal forces *are* affected *by relative cross-sectional dimensions of structural members and by relative Young's modulus.*

Assumption 2.8 (Statically Indeterminate Truss II).
Internal forces *are* significantly affected *by a slight lack of fit of a structural member or support displacement.*

Euler Buckling Constraints for Local Stability
Local instability of a structural member occurs when the compression force exceeds its critical buckling load (Tyas et al. 2006). A well established method to ensure local stability is to define the stability limit of an axially loaded prismatic structural member using the concept of *Euler's critical load* (Timoshenko et al. 1962), even though the concept may overestimate the actual buckling strength (Descamps and Coelho 2014). This concept provides a criterion for members made of one material satisfying Hooke's law (Rychlewski 1984). A structural member is said to be a *perfect structural member* if it satisfies the previous conditions. The principal significance of a perfect structural member is that it allows us to preprocess the

upper bound of the capacity of a structural member, see Section 5.2. If the structural member does not deflect laterally, the corresponding state of equilibrium is said to be a *trivial state of equilibrium*. Exceeding Euler's critical load, multiple unstable states of equilibrium can arise.

Let us introduce the temporary notation F_b, l_b, σ_b, s_b for Euler's critical load, critical lateral deflection (buckling) length, critical stress, and the safety factor against buckling of an axially loaded prismatic structural member b with uniform axial stress, respectively. By assuming the bending stiffness $EI = const$, where I_{min} is the minimal area moment of inertia (Gibson 2016) of a structural member having the cross-section Ω, the following is valid

$$F_b = \frac{\pi^2 E I_{min}}{l_b^2}, \qquad (2.29)$$

$$\sigma_b = \frac{F_b}{\Omega} = \frac{\pi^2 E I_{min}}{l_b^2 \Omega}, \qquad (2.30)$$

$$s_b = \frac{F_b}{F}, \qquad (2.31)$$

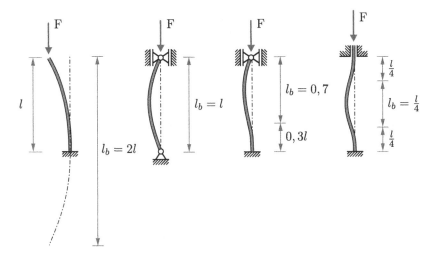

Figure 2.11 Critical buckling lengths for Euler's critical load: (left) lower end fixed for rotation and translation, upper end free for rotation and translation; (middle left) both ends free for rotation and fixed for translation; (middle right) lower end fixed for rotation and translation, upper end free for rotation and fixed for translation; (right) both ends fixed for rotation and translation. (Obtained from Mang and Hofstetter 2013)

where F is an external force. Different critical buckling lengths l_b for Euler's critical load are illustrated in Figure 2.11, with l being the original length of a structural member. As mentioned by Rozvany (1996a), the implementation of Euler buckling constraints for local stability will create a deviation from the optimum topology. The reason thereof is simply that if the critical buckling length and the corresponding slenderness ratio increase, the critical stress decreases and the required cross-sectional area increases.

2.2 Additive Manufacturing

This section provides an overview of AM. A very comprehensive overview of AM can be found in Bikas et al. (2016), Frazier (2014), Gardan (2016), and therefore only essential properties of AM technologies are presented in this section. Further, as a part of this thesis deals with the optimization of self-supporting truss-like structures, an emphasis is placed on the subject area, support structures.

2.2.1 Basics

The term AM, in public communication usually less precisely referred to as 3D Printing, defines an additive process for rapid form manufacturing, where the final object is created "from scratch" by adding material in layers; each layer is a thin cross-section of the component derived from the original 3D-CAD data (Burns 1993, Gibson et al. 2014). Researchers at the Massachusetts Institute of Technology invented an ink-jet printing-based technology and coined the term 3D Printing, which is based on the technological similarities to two-dimensional ink-jet printing (Sachs et al. 1990). AM was, initially, and still is in some industrial sectors, referred to as rapid prototyping (Hopkinson and Dicknes 2003). Nowadays, the term rapid prototyping seems to be outdated and indicates rather the historical development of AM, since the term rapid prototyping was used as an umbrella term to define technologies quickly manufacturing physical prototypes from 3D-CAD data instead of the end-of-use product manufacturing process which is AM nowadays (Gibson et al. 2014). Over the last four decades, since Hull (1984) invented StereoLithogrAphy (SLA), the first commercial AM technology, a range of different AM technologies have been enhanced to such consistent manufacturing quality that AM has become relevant for functional end-use components. For these reasons, AM has been used in the automotive, mechanical engineering, and aerospace sectors (see, e.g., Leal et al. 2017, Liu et al. 2017, Marchesi et al. 2015, and the references therein). An example AM method is the so-called group of solid (powder)-based[5] AM systems, such as Selective Laser Sintering (SLS), which uses, e.g., a laser to selectively fuse a layer-by-layer structure composed of a polymer or metal powder in a build chamber of an AM machine. SLS shows how AM differs from conventional manufacturing processes such as subtractive processes (e.g., milling, cutting, or drilling), formative processes (e.g., casting or forging), and joining processes (e.g., welding or fastening), see Conner et al. (2014).

In recent years, there have been a series of commercial hybrid manufacturing processes combining additive with conventional, in particular subtractive, manufacturing methods performed on one hybrid manufacturing platform (Grzesik 2008, Zhu et al. 2013). The goal is to improve the finish surface roughness by avoiding multiple manual post-processing activities, which would be necessary if only AM were to be used. Figure 2.12 illustrates AM's fundamental concept, consisting of three steps: First, a digital CAD data (3D object model) of the designed component is generated. Second, using a so-called slicing software (see, e.g., Gebhardt 2011) and a water-tight triangular mesh (see, e.g., Gao et al. 2015) for the boundary/surface of the 3D object model, the object gets divided into a batch of flat layers by intersecting multiple horizontal planes, which are manufacturable by linear movements (toolpaths) of an AM machine (Gao et al. 2015). Third, layerwise deposition is used to convert the digital 3D object model into a physical component.

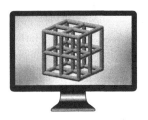  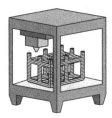

Figure 2.12 The fundamental concept of AM: (left) generation of a 3D object model and water-tight triangular mesh of the surface of the 3D object model; (middle) slicing software dividing the 3D object model into a batch of flat layers and generating toolpaths of an AM machine; (right) layerwise deposition is used to convert the digital 3D object model into a physical component. Note that the machine type shown is only symbolic and does not satisfy the range of AM processes according to ASTM F42—Additive Manufacturing standards

---

[5] Solid-based AM systems that use powder as starting material form the subclass powder-based AM systems.

The staircase effect, the ability to accurately reproduce a given layer, the quality of the bonding of layers, and the material microstructure are the main factors the quality of an additively manufactured component depends on (Gibson et al. 2014, Yasa et al. 2016). AM's primary materials are various ceramics, metals, and polymers, partially as high-performance materials suitable for end-use components (Ngo et al. 2018). For a better comparison of the different materials for AM, please refer to Table A.1 in the electronic supplementary material on page 2. An assorted selection of standard materials for AM (see also Bourell et al. 2017, Gibson et al. 2010, andthereferencestherein) is shown.

AM is ideal for individual customization and manufacturing of components from batch size one, since the direct unit cost is virtually independent of the number of units (Clausen 2016). In contrast, traditional manufacturing methods, e.g., injection molding, are best for large batches of (solid) end-use components without manual post-processing activities, to compensate for high tool costs (Dewhurst and Boothroyd 1988, Chen and Liu 1999). An advantage of AM is that it provides high design and geometric freedom (Huang et al. 2013), reduction of storage costs (Khajavi et al. 2014), and mass customization of components (Kwok et al. 2017). Next, geometric complexity is claimed to be free, as the component is manufactured layerwise, such that costs and manufacturing time are independent of the component complexity (Atzeni and Salmi 2012). Besides, one essential benefit of AM is that it implies the possibility of reducing the tooling time or eliminating it altogether. Next, the number of work and assembly steps and devices can be reduced. Apart from that, AM is a key enabling technology for lightweight design of shape and topologically optimized components, since it enables the manufacture of these components, unlike most traditional manufacturing methods. Additionally, AM enables part (component) count reduction through exploiting component consolidation, function integration, and optimization (see, e.g., Thompson et al. 2016). Even parts with high design and geometric complexity can be manufactured in one step, which implies eliminating manufacturing steps.

One disadvantage of AM is its limited suitability for industrial mass production due to high manufacturing costs resulting from low build rates (Thomas and Gilbert 2014). Besides, AM is usually not scalable due to the discontinuous manufacturing process, which prevents economies of scale. Another weakness, depending on the process methodology, is obligatory post-processing for end-use components due to possible lack of precision in manufacturing, surface roughness, and component anisotropy. It is a well-known fact that additively manufactured components exhibit anisotropy, caused largely by the orientation of the AM system and the process-related joining[6] of the layers (see Carroll et al. 2015, Kok et al. 2018, Popovich et al. 2017, and the references therein). These anisotropic material properties are hard to predict exactly (see, e.g., Ahn et al. 2002 Ning et al. 2015), which renders the simulation of component properties significantly more computationally intensive (Gao et al. 2015). As described by Gibson et al. (2014), anisotropy can be eliminated up to a certain extent, but not completely, by performing a final heat treatment. For some process methodologies, dimensional accuracy and surface finish of AM are, because of the nature of the layer-by-layer process, substandard compared to traditional manufacturing methods. Moreover, the component size is often subject to the build chamber size (see, e.g., Gebhardt 2011), which excludes large component applications.

2.2.2 Technologies

Various grading techniques can be found, in the field of AM. As mentioned by Burns (1993), Kruth et al. (1998), one approach is to categorize AM technologies according to the baseline technology; a second approach is to categorize according to the starting material (Chua et al. 2010). One of the major drawbacks of using these one-dimensional grading techniques is that a delusive allocation of machine types can occur, e.g., grouping SLS with 3D Printing (Gibson et al. 2014). A very plausible two-dimensional grading technique[7] was stated by Pham and Gault (1998), endorsed by experience. Gibson et al. (2014) and Clausen (2016) have adapted Pham and Gault's grading technique to recent AM technologies and standard ASTMF2792-12a (ASTM terminology standard).

In order to state which AM machine types have processing similarities in common, we briefly review the seven process categories for AM defined in standard ASTMF2792-12a in the following paragraphs. That is, we can introduce a category of machines, rather than needing to state each single variation of a machine type. For a general and comprehensive review of AM technologies, please refer to Bikas et al. (2016), Chua and Leong (2014), Frazier (2014), Gardan (2016), Hopkinson et al. (2006) and the references therein.

---

[6] Generally speaking, the material properties are worse in build direction than in the respective layer, in layered manufacturing.

[7] First dimension: Generalized channel setup of the AM method; Second dimension: starting materials by physical form.

Directed Energy Deposition Directed energy deposition methods, also referred to as Laser Engineered Net Shaping (Atwood et al. 1998), Direct Metal Deposition (Lewis and Schlienger 2000), or Electron Beam Additive Manufacturing (Gong et al. 2012), simultaneously deposit material and energy, using, inter alia, orifices or nozzles. Similar to extrusion methods, spatial structures are created by selective deposition of material layer by layer. The process methodologies were originally used for coating and component repair of, e.g., engine combustion chambers, airfoils, blade integrated disks (blisks), or turbine blades (Portolés et al. 2016). Throughout the selective deposition, no uncured powder exists at the end of the manufacturing process. If the build platform is pivoted, i.e., implemented in a multiaxial machine providing 3D positioning, the possibility to manufacture without a support structure exists. Directed energy deposition processes are sufficient for manufacturing components with material properties at least equivalent to standards achievable with conventional manufacturing processes. A subset of directed energy deposition processes has no build chamber, and consequently no space limitations, making them suitable for large component applications. A serious disadvantage is limited precision, both in terms of the minimum feature size and geometrical deviations. Therefore, directed energy deposition methods are less suitable for complex components than powder-based AM systems; however, wire-feed laser and arc beam deposition processes are suitable for aerospace applications (Brandl et al. 2010). Post-treatment of the components is obligatory (Gibson et al. 2015, Saboori et al. 2017).

Material Extrusion Material extrusion can be categorized into fused filament (Brenken et al. 2018) and direct ink writing (Lewis 2006) processes and is currently the most popular AM process technology on the market (Campbell et al. 2018). In the fused filament processes, thermoplastics are partially melted using a heated nozzle and deposited linearly and layerwise. The heated thermoplastic melts the previous layer slightly, cools down through thermal conduction, and a rapid solidification process begins. A support structure is necessary. It either consists of the base material and must be removed mechanically or consists of water-soluble material. In direct ink writing, a liquid-phase ink is extruded pneumatically, driven using a nozzle. For the reasons stated, post-curing is obligatory. The major advantage of material extrusion processes, especially Fused Layer Modeling and Robocasting, is the technically simple, and therefore economically attractive, process. Besides the industrial applications of AM, mainly inexpensive machines for home users or office environments are available on the market. Extrusion processes are partially sufficient for manufacturing components with material properties ready for series production and at least equivalent to standards achievable with injection molded parts. The range of materials usable is relatively large, so that a change of material is easy. Parallel processing of multiple materials is realized by using multiple nozzles. Manual mechanical removal of the support structure can be avoided by using, e.g., water-soluble support material. The major disadvantage of extrusion processes is the technological necessity of support structures, which makes post-processing expensive. Furthermore, the dimensional and surface accuracy (layer thickness, surface quality, and roughness) is determined by the nozzle.

Material Jetting Material jetting (Loh et al. 2018) is a technology related to two-dimensional ink-jet printing but solidifies photopolymer materials in a single step using ultraviolet light for polymerization. The base material (photopolymer) is deposited in drops using a print head with multiple nozzles and immediately cured using an on-board ultraviolet lamp (Barclift and Williams 2012, Gay et al. 2015). A low melting point wax-like thermoplastic structure cured simultaneously with the base material or a needle-like slender structure of the base material (photopolymer), extruded through at least one nozzle, is used as support. During post-processing, the support material can be easily heated, melted, and removed once the base material has cured. Material jetting has the advantage of high manufacturing precision and is suitable for manufacturing complex or thin-walled or both types of parts with high surface quality. Due to the low processing temperature of the photopolymer materials, shrinkage and warpage are low. Two main disadvantages of material jetting are the technological limitations of photosensitive materials, which are relatively expensive, and the layerwise process is rather slow. Besides, a limitation also exists of photopolymer materials being expensive, as well as the need for a support structure. Due to poor physical and mechanical properties, the process methodology is not suitable for functional parts.

Powder Bed Fusion Powder bed fusion (Gibson et al. 2014) is a layerwise AM process of repeatedly adding layers of polymer or metal powder in a build chamber. The powder bed gets selectively fused using an electron beam, thermal printhead, or laser. Many industrially-suited AM systems belong to this category. Unlike conventional sintering methods, powder bed fusion does not need long diffusion time or high pressure. The manufacturing process takes place in a build chamber filled with high-purity inert gases, e.g., nitrogen or argon, to avoid atmospheric impurities. Uncured but used powder material enclosing the cured component structure functions as support material and can be partially recycled, mixing used powder and virgin powder into a new powder blend. The main advantage is the wide variety of materials. The material properties are very good for both plastic and metal components, especially for the latter, and the material properties can approach the properties of conventionally manufactured components. For the reasons stated, no support structures are required for plastics.

A serious disadvantage is that the component surface has high roughness values, depending on the closeness of grain, so that time-consuming post-processing, e.g., sanding, is necessary. Especially in the case of internal cavities and thin channels, the cleaning of the components is an elaborate post-processing step. Powder bed fusion is a complex manufacturing process, involving high investment and operational costs. The heating and cooling process is tedious and influences the component's lead time negatively. Material changes are cost-intensive due to the necessary cleaning work. The basic system design of powder bed fusion is shown in Figure 2.13 using the manufacturing technique SLS.

Sheet Lamination Sheet lamination (Gebhardt 2011), also referred to as Layer Laminate Manufacturing[8], is a subtractive-additive hybrid process in which a sheet of metal, polymer, or paper is bonded layerwise (Gibson et al. 2014). Before each layer build-up, the material is cut to the component contour, performed via a blade cutting, laser, or milling cutting system. The bonding process is achieved by means of polymerization using the bonding of structural adhesives or welding. The order of cutting and bonding is invertible. The sheet lamination processes offer a number of advantages. First, the technically straightforward process permits the utilization of a wide range of materials. Second, the process is suitable for rapidly manufacturing solid or large or both types of components. The major disadvantage is low material properties. Besides, the removal of material, especially for internal cavities, is complex or even impossible. For this reason, manufacturability is constrained to non-complex component geometry. Post-processing is obligatory and, in contrast to other AM methods, expensive.

Vat Photopolymerization Vat photopolymerization (Gebhardt 2011) selectively cures a liquid or paste-like photosensitive resin, i.e., a photopolymer stored in a vat, point-by-point using ultraviolet light. A well-established method is to use a laser and scanner combination (Gibson et al. 2014). The ultimate benefit of SLA (Gebhardt 2011), the first[9] and most widely used vat photopolymerization tech-

---

[8] Occasionally, the protected product name Laminated Object Manufacturing is used for this group of methods.

[9] In 1987, Co-Founder and Chief Technology Officer of 3D Systems, Charles (Chuck) Hull, commercialized SLA by inventing the first-ever 3D printer SLA-1. In 2016, the SLA-1 was designated a Historic Mechanical Engineering Landmark by the American Society of Mechanical Engineers (ASME).

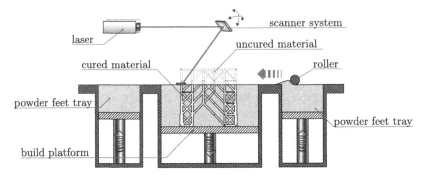

Figure 2.13 The basic principle of SLS: Using a laser and a scanner system, the powdered material is sintered layerwise into a solid structure based on a 3D object model. Each layer of powder is deposited on the build tray and leveled using a roller. After sintering a solid layer using the laser, the build tray or powder feed tray moves downwards or upwards such that a new layer of powder can be deposited and sintered. This process is continued until the component is finished. See Figure 7.2b for the component manufactured by SLS as shown in the sketch. (Obtained from Gibson et al. 2014)

nique, is the highest manufacturing precision among all AM processes. Besides, complex or thin-walled or both types of components with a high surface quality are manufacturable. Depending upon the machine specifications, e.g., using a laser and scanner combination, the uncured resin can be recycled. The manufacturing precision is mainly limited by the laser beam's diameter and not by the physical limits of the manufacturing process. Shrinkage and warpage are low, due to the low processing temperature of the photopolymers. Two main disadvantages of vat photopolymerization are the limitation of photosensitive materials being relatively expensive and the rather slow layerwise manufacturing process. Besides, depending on the machine type, internal cavities can only be emptied and finished through drainage openings, and post-curing using ultraviolet light might be necessary. The material costs of the filament are relatively high compared to the granules used for injection molding. A support structure for overhangs and undercuts is obligatory.

Binder Jetting Binder jetting (Gebhardt 2011), also called 3D Printing, is a layerwise AM process combining powder bed and jetting methods. A liquid bonding agent with varying properties is selectively deposited to join material in a powder bed. The uncured material enclosing the cured component structure serves as a support for, e.g., overhangs and undercuts, and fixes the component location in relation to the machine coordinate system. Thus, a key benefit of binder jetting is that no support structure is necessary. At the end of the process the uncured powder material gets vacuumed off. A common post-treatment method is sintering. Two main disadvantages of binder jetting are rather low strength properties and low surface accuracy.

2.2.3 Support Structures

Although AM offers high design freedom and geometric complexity is of minor importance, it also implies manufacturing constraints, such as a minimum slot and wall thickness, a minimum size between features, and a constraint on the inclination of downward-facing surfaces of a component, the so-called overhang limitation (Thomas 2009). Thus, components exhibiting a certain degree of geometric complexity require so-called support structures, i.e., auxiliary structures not part of the component, to keep cured material in the designated position and tethered to the build platform in the course of manufacturing (Jiang et al. 2018). As can be seen in Figure 2.14, the angle between a downward-facing surface and the build platform, the so-called overhang angle δ, is decisive for the manufacturability of the component. Let δ_m be the minimum self-supporting and manufacturable overhang angle. The overhang limitation has been intensively studied for various process methodologies (see, e.g., Kranz et al. 2015, Mertens et al. 2014), with the result that the minimum overhang angle varies for different process conditions (see, e.g., Cloots et al. 2013, Wang et al. 2013) but typically amounts to $\delta_m = 45°$ for Selective Laser

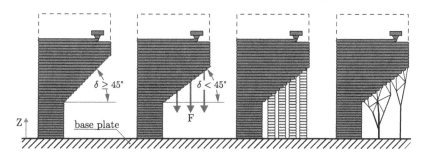

Figure 2.14 Example component manufactured with SLM and build direction Z: (left) self-supporting component design by using $\delta_m = 45°$; (middle left) not manufacturable component design by using $\delta < \delta_m$; (middle right) manufacturable component supported by unoptimized regular lattice structure; (right) manufacturable component supported by optimized truss-like structures

Melting (SLM) (Langelaar 2016). As detailed in Figure 2.15, depending upon the (thermo-) mechanical processes of the AM technique, the need for support structures can also arise due to the stresses inherent within the manufacturing process, shape distortion, inhibit deflection, or to balance the component (Zhang and Chou 2008).

For material extrusion methods, material jetting methods, and directed energy deposition methods, an underlying layer[10] is mandatory to deposit material, due to the fact that a layer needs thermal conduction and mechanical support from the previously built layer (van de Ven et al. 2018). This limits the overhang of a layer to the layer underneath. Material extrusion methods require protection against gravity. For material jetting and direct energy deposition methods, the layer-based process requires a previous substrate layer to be built up. For metal AM processes, e.g., directed energy deposition and powder bed fusion, support structures are primarily driven by anchoring towards thermally induced residual stresses (Withers and Bhadeshia 2001a,b), since material stiffness and temperature gradients are several orders of magnitude higher than for polymers (Clausen 2016). Consequently, gravity-based support structures are of minor importance in metal AM processes. For a detailed study on residual stresses in SLS and SLM, please refer to Mercelis and Kruth (2006) and the references therein. Vat polymerization and powder bed AM processes are, due the nature of the layer-by-layer process, limited such that the support structures are made of the same material as the main component. In contrast, some material extrusion or material jetting processes allow multi-material support (see, e.g., Bandyopadhyay and Heer 2018, Priedeman Jr and Brosch 2004). Support structures anchoring towards thermally induced residual stresses for binder jetting and non-metal powder bed fusion processes can be avoided by using a preheating system, making the temperature gradients minuscule. In addition, the uncured material enclosing the cured component structure serves as support.

The support structures have to be considered in the design process and removed in an expensive, time-consuming, and non-destructive post-processing step, e.g., cutting, milling, or using solvents. To ensure that all surfaces, especially internal structures, are accessible during post-processing for the machining tool and all support structures can be removed, the component must be correctly designed and oriented during manufacturing. The support structures lead to time, energy, and material waste as well as higher build time (Jiang et al. 2018). In addition, the finished surface roughness in the connecting areas between the support structure and the component is reduced. Especially for metal parts, the component's functional surface must be mechanically finished, meeting the required tolerances, dimensions, and

---

[10] The first layer, the so-called bottom layer, connects the build platform and the component.

Figure 2.15 (Left) Powder bed manufactured airbrake hinge bracket inclusive of support structures anchoring towards thermally induced residual stresses and tethering the component to the build platform. The hole is machined. (Right) Two final airbrake hinge brackets after post-processing and support removal. Both figures from (Hamilton 2016), reproduced with permission from the author and the publisher, © 2016 Inovar Communications Ltd., Shrewsbury, UK

surface requirements to guarantee component function. This post-processing is often carried out using the same methods as for series cast components, and requires hand or tool access or both. Consequently, optimizing or eliminating the support structures is motivated by significant economic interest and manufacturing advantages. The basic idea of optimizing support fixing against gravity is explained using the example displayed in Figure 2.14. To minimize waste, material consumption, energy, and cost of AM, and thus to make further improvements in AM compared with traditional manufacturing, research in support structures is of great importance. This topic is dealt with in greater detail in Sections 4.5, 5.3 and 7.2.

2.3 Mathematical Programming

Considering manufacturability, design freedom, and post-processing effort of AM, the combination of global optimization and topology optimization, especially MILP and TTO, can be used to design periodic or stochastic high-strength truss-like structures or both for AM (Achtziger and Stolpe 2008, 2009). Therefore, many distinct directions for further research exist, with one of them being large-scale optimization of additively manufactured truss-like structures based on the high performance of commercial linear programming software like CPLEX and GUROBI (GUROBI 2019), to solve large-scale real-world TTO problems. To implement engineering (structural) design rules and goals as a mathematical optimization program, we need several essential mathematical preliminaries, which are explained in this section. General references are Schrijver (1998), Suhl and Mellouli (2009) and the references therein.

2.3.1 Graph Theory and Flow in Networks

To facilitate access to graph-based TTO, see Chapters 4 and 5, we give a concise but self-contained introduction to graph theory and flow in networks. Network flow problems and the corresponding (linear) models are far-reaching areas within OR (Bazaraa et al. 2011). The foundation of the short introduction to the graph theory and flow in networks is based on Diestel (2006) and Schelbert (2015).

Graph Theory

An *undirected graph* $G := (V, E)$ is a tuple consisting of two finite sets; the set of nodes (vertices) $V \neq \emptyset$ and the set of edges E. With $u, v \in V$ and $\{u, v\} = e$ as an (unordered) set of two nodes, we can define an edge $e \in E$. If an edge $e = \{n_1, n_2\}$ connects two nodes n_1, n_2, these nodes are defined to be *adjacent*. A node $v \in V$ is *incident* with an edge $e \in E$ if $v \in e$. An edge $e = \{n_1, n_2\}$ with $n_1 = n_2$ is called a *self-loop*. To exclude self-loops one can assume $n_1 \neq n_2$.

A *directed graph (digraph)* $G := (V, A)$ is an orientation of an undirected graph G by directing one of its nodes to the other node and is given by two sets; the set of nodes V and the set of *arcs* A. We define an arc $a \in A$ to be a *directed edge*. More precisely, an arc $a \in A$ is an ordered tuple (n_1, n_2), where n_1 is called the *tail* and n_2 the *head* of the arc.

A *weighted graph* $G := (V, E, c)$ demands an additional cost function $c : E \to \mathbb{R}$ defining the individual costs of each edge $e \in E$. A directed or undirected *multigraph* is a pair $G := (V, E)$ of nodes and edges, with the map $E \to V \cup V^2$ assigning one or two nodes to every edge. Since an edge $e = xy$ with $x \neq y$, indicating the arcs between x and y, is not identified uniquely by the pair (x, y) or (y, x), we define a set of directed edges as triples

$$\vec{E} := \{(e, x, y) \mid e \in E, \ x, y \in V, \ e = xy\}. \tag{2.32}$$

Having a subset $S \subseteq V$ of nodes, we can define the incident edges and edges inside the subset

$$\gamma(S) := \{\{n_1, n_2\} \in E : n_1 \in S\}, \tag{2.33}$$

$$E(S) := \{\{n_1, n_2\} \in E : n_1, \ n_2 \in S\}. \tag{2.34}$$

Define $\gamma^+(S), \gamma^-(S)$ for a directed graph, where we have a set for the incoming and a set for the outgoing arcs of the subset $S \subseteq V$, respectively. For abbreviation, we use $\gamma^+(n), \gamma^-(n)$ for a single node n. It follows that

$$A(S) := \{(n_1, n_2) \in A : n_1, \ n_2 \in S\}, \tag{2.35}$$

$$\gamma^-(S) := \{(n_1, n_2) \in A : n_1 \in S\}, \tag{2.36}$$

$$\gamma^+(S) := \{(n_2, n_1) \in A : n_1 \in S\}. \tag{2.37}$$

Flow Networks

Let a *network* N be given by the tuple $N := (G, s, t, c)$, where we have a multigraph $G := (V, E)$, two fixed nodes $s, t \in V$, and the *capacity function* $c : \vec{E} \to \mathbb{N}$ on G. We define the function $f : \vec{E} \to \mathbb{R}_+$ to be a non-negative *flow* of the *flow graph network* N if the following three conditions are satisfied:

$$f(e, x, y) = -f(e, y, x) \quad \forall (e, y, x) \in \vec{E}, \ x \neq y, \tag{2.38}$$

$$\sum_{e \in \gamma^-(v)} f(e) = \sum_{e \in \gamma^+(v)} f(e) \quad \forall v \in V \setminus \{s, t\}, \tag{2.39}$$

$$f(\vec{e}) \leq c(\vec{e}) \quad \forall \vec{e} \in \vec{E}. \tag{2.40}$$

For definiteness, Equation (2.38) defines the *net flow* of a directed edge $\vec{e} \in \vec{E}$, known as *skew symmetry*. Equation (2.39) forces the net flow to be zero at each node $v \in V$, excluding the *source* s and *sink* t ($s \neq t$), called *flow conservation*. Equation (2.40) ensures that the flow of a directed edge cannot exceed its non-negative capacity c, called *capacity constraints*.

A *supersource* s or *supersink* t or both can be added to the network to act as a source or sink for each of the original sources and sinks. The task is now to construct

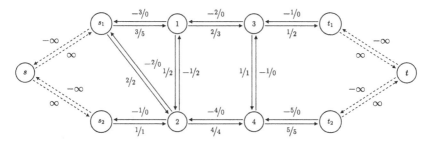

Figure 2.16 A plane flow network with sources s_1, s_2, sinks t_1, t_2, supersource s, and supersink t. The flow and capacity is denoted as f/c. Note that we use dashed lines for the directed edges linking the supersource and supersink to the network, and require that the flow network applies with skew symmetry, flow conservation, and capacity constraints

a flow network with a single supersource or supersink or both, see dashed arcs in Figure 2.16, out of a flow network with multiple sources and sinks.

Klarbring et al. (2003) developed a new method for topology optimization of flow networks and Evgrafov (2006) investigated a simultaneous optimization of topology and geometry of flow networks. Both authors conclude that the methodologies of flow networks and TTO are mutually beneficial. Also, our models (see Chapters 5 and 7) can be interpreted (in general) as flow networks with sources as external forces, sinks as bearings, and capacities (upper bounds) of structural members.

2.3.2 Mixed-Integer Linear Programming

A MILP is a mathematical optimization program describing a mathematical optimization problem. More precisely, a MILP consists of two sets: A set of variables and a set of linear constraints on these variables. The set of linear constraints includes linear equations and inequalities. The set of variables consists of variables having a restriction to integer values or real values. The linear objective function gives the quantity to be optimized. The goal of the optimization is to find a variable assignment that satisfies all constraints and minimizes or maximizes the given linear objective function.

A mathematical optimization program is said to be a MILP if the objective function and all constraints are linear, and a Mixed-Integer Nonlinear Programming (MINLP) if the objective function or at least one constraint is nonlinear. Our previously introduced assumptions and the case of a minimization problem lead to the formal notation of a Mixed-Integer Programming (MIP)

$$\min \quad c^T x$$
$$\text{s.t.} \quad Ax \le b \tag{2.41}$$
$$x \in \mathbb{Z}^{PG} \times \mathbb{R}^{n-PG}$$

where $A \in \mathbb{Q}^{m \times n}$, $b \in \mathbb{Q}^m$, $c \in \mathbb{Q}^n$ with PG, n, $m \in \mathbb{N}$, $PG \le n$ is the instance data. There exist PG integer and $n - PG$ continuous variables. It is sufficient to show the standard minimization problem, since it can be converted into the associated standard maximization problem by a trivial multiplication of the objective function by -1. A feasible solution for problem (2.41) exists if the vector $x' \in \mathbb{Z}^{PG} \times \mathbb{R}^{n-PG}$ is subject to the condition

$$Ax' \le b. \tag{2.42}$$

Claiming a *global optimum*

$$x', c^T x' \leq c^T x \qquad (2.43)$$

holds for all feasible solutions x. We can identify multiple types of mathematical optimization programs depending on the characteristics of PG, n, and x. A MILP with $PG = 0$ is said to be LP. Analogously, a MILP with $PG = n$ integer variables is said to be an Integer Programming (IP). A MILP with $PG = n$, $x \in \{0, 1\}$ is a special case called binary program.

2.3.3 The Optimality Gap

Given a feasible solution as vector $x \in X$ for the MILP presented in Constraints (2.41) it is not ensured that the feasible solution is a global optimum. Since it is not always possible to solve a mathematical optimization problem to global optimality, e.g., because of a reasonable time, the natural question arises whether it is possible to determine the gap between the incumbent best-known solution and the global optimal solution. Against this backdrop, the primary strategy is to bound the global optimal solution using a *primal bound* and *dual bound*, deriving a so-called *duality gap*. Assuming a minimization problem, the primal and dual bounds are called *upper bound* and *lower bound*.

To be specific, consider a minimization problem and a feasible solution as vector $x \in X$ providing the (primal) upper bound with

$$z_{\text{primal}} = c^T x \qquad (2.44)$$

for the optimal objective function. The advantage of using *primal methods* lies in the fact that one can determine feasible solutions computationally efficient and fast, whereby standard fields of application are simulated annealing, tabu search, and meta-heuristics like genetic algorithms (Altherr 2016). The disadvantage is that, without concurrent use of primal and *dual methods*, in general it is not possible to derive the size of the optimality gap and consequently to guarantee global optimality. One measure of the optimality gap is the duality gap.

By dropping hard constraints of the MILP (2.41), we can relax the problem and solve the *relaxed problem*, which is easier to solve and whose feasible set contains every feasible solution for the original problem. The procedure is to find lower (dual) bounds. A mathematical optimization problem OP in the form

$$z^* = \min\{g(x) : x \in X \subseteq \mathbb{R}^n\} \tag{2.45}$$

allows the easier relaxed problem OP' in the form

$$z' = \min\{g(x) : x \in T \subseteq \mathbb{R}^n\}, \tag{2.46}$$

satisfying

$$X \subseteq T, \tag{2.47}$$

with the following properties:

- If $x' \in X$ is an optimal solution for problem OP' and x' is a feasible solution for OP, then x' is also the optimal solution for OP.
- If OP' is infeasible, OP also is infeasible.
- $z' \leq z^*$.

We conclude that z_{dual} (z_{primal}) is the objective value of each dual (primal) solution and is a lower (upper) bound to the global optimal solution z^*, with the property that

$$z_{\text{dual}} \leq z^* \leq z_{\text{primal}}, \tag{2.48}$$

and finally we can state the *optimality gap*

$$g := \frac{z_{\text{primal}} - z_{\text{dual}}}{z_{\text{dual}}}. \tag{2.49}$$

The optimality gap provides the (relative) difference between the primal and dual bound, whereas if $g = 0$ and thus $z_{\text{dual}} = z' = z_{\text{primal}}$, then z' is an optimal solution. If there exists a strictly positive gap g for a solution for an optimization problem OP, we would have no proof that the solution for OP is optimal and the knowledge of the gap g is at best an indicator for the worst-case deviation, instead of the actual deviation from the global optimum. The standard procedure is to find both bounds until g is approximately zero, where equality holds if the solution is a global optimum.

Quantified Mixed-Integer Linear Programming
QMIP is a direct and formal extension to MILP utilizing uncertainty bits. In QMIPs the variables are ordered explicitly and they are quantified either existentially or universally resulting in a multistage optimization problem under uncertainty:

Definition 2.4 (Quantified Mixed-Integer Linear Program)
Let there be a vector of n variables $\mathbf{x} = (x_1, \ldots, x_n)^\top \in \mathbb{Q}^n$, lower and upper bounds $\ell, \mathbf{u} \in \mathbb{Q}^n$ with $\ell_i \leq x_i \leq u_i$, a coefficient matrix $A \in \mathbb{Q}^{m \times n}$, a right-hand side vector $\mathbf{b} \in \mathbb{Q}^m$ and a vector of quantifiers $\mathbf{Q} = (Q_1, \ldots, Q_n)^\top \in \{\forall, \exists\}^n$. Let $I \subset \{1, \ldots, n\}$ be the set of integer variables and $\mathcal{L}_i = \{x \in \mathbb{Q} \mid (l_i \leq x \leq u_i) \wedge (i \in I \Rightarrow x \in \mathbb{Z})\}$ the domain of variable x_i and let $\mathcal{L} = \{\mathbf{x} \in \mathbb{Q}^n \mid x_i \in \mathcal{L}_i\}$ be the domain of the entire variable vector. The term $\mathbf{Q} \circ \mathbf{x} \in \mathcal{L}$ with the component wise binding operator $\circ$ denotes the quantification sequence $Q_1 x_1 \in \mathcal{L}_1, \ldots, Q_n x_n \in \mathcal{L}_n^\top$ *such that every quantifier Q_i binds variable x_i ranging in its domain $\mathcal{L}_i$. We call*

$$z = \min c^\top x$$
$$s.t. \quad \mathbf{Q} \circ \mathbf{x} \in \mathcal{L} : A\mathbf{x} \leq \mathbf{b}$$

a quantified mixed-integer linear program (QMIP).

Note that the objective function is actually a *minmax function* alternating according to the quantification sequence: Existential variables are set with the goal of minimizing the objective value while obeying the constraint system whereas universal variables aim at a maximized objective value. For more details, please refer to Ederer et al. (2011), Hartisch (2020), Lorenz and Wolf (2015), Subramani (2003, 2004).

A solution is a strategy for assigning existentially quantified variables such that the linear constraint system

$$A\mathbf{x} \leq \mathbf{b} \tag{2.50}$$

is fulfilled. In particular, in a solution it is ensured that even for the worst-case assignment of universal variables $A\mathbf{x} \leq \mathbf{b}$ holds. For our computational experiments concerning our QMIP $\text{TTO}_{1;q}$ (see Section 7.4) we create the corresponding Deterministic Equivalent Problem (DEP) (Wolf 2015) and solve the resulting MIP using CPLEX. The QMIP $\text{TTO}_{1;q}$ presented in this work features two quantifier changes in the quantification sequence. Therefore, it can be interpreted as an adjustable robust mathematical optimization problem with right-hand side uncertainty (Yanıkoğlu et al. 2019).

Optimization of Truss Structures

3

In this chapter, we introduce the optimization of truss structures. In Section 3.1, we define the problem statement of truss optimization used to find conceptual and preliminary designs for further detailed analysis and redesign. Section 3.2 provides a two-tier overview of related work, seen from the perspective of global optimization as well as engineering.

3.1 The Problem Statement

The purpose of this section is to provide a detailed overview of truss optimization. We first introduce the theoretical basics of truss optimization before we discuss and contrast complexity in engineering with computational complexity and symmetry in structural mechanics with symmetry as a property of optimality in mathematical design optimization. Besides, we introduce the notion of the ground structure approach. We complete this section with a compact overview of the general form of a TTO problem.

3.1.1 Theoretical Basics

The Main Idea
In the field of solid mechanics and its applications, truss optimization is a subfield of structural optimization, whereas the latter is a subfield of mathematical design optimization (Christensen and Klarbring 2008). Weight, critical load, state of stress, element stresses, stiffness, nodal displacements, geometry, natural frequencies, and buckling of the truss are among the most commonly investigated types of objective functions in truss optimization (Tejani et al. 2018a). According to Christensen and

© The Author(s), under exclusive license to Springer Fachmedien Wiesbaden GmbH, part of Springer Nature 2022
C. Reintjes, *Algorithm-Driven Truss Topology Optimization for Additive Manufacturing*, https://doi.org/10.1007/978-3-658-36211-9_3

Klarbring (2008), the decision variables are constrained to the quantities stresses, displacements, or the geometry, whereas the quantities may also be used as objective functions. The pure mechanical structural performance of a feasible solution is measured by the objective function, neglecting functionality, economy, or esthetics in the general case (Christensen and Klarbring 2008) (see, e.g., Figure 3.1 for the basic principle of a truss optimization problem).

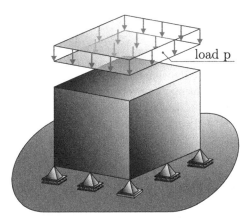

Figure 3.1 A general truss optimization problem. Find the optimal design of the load-carrying truss structure, given a predefined design domain. Note that the bearings shown are only symbolic

Three Types of Truss Optimization
In their review of truss optimization, Tejani et al. (2018a) state that truss optimization has become a fast-emerging research area since the last three decades. Reviewed in a structural design setting, three categories depending on the design variables exist, specifically: sizing optimization, also known as structural thickness (cross-sectional) optimization, shape optimization, also known as geometry optimization, and TTO, also known as layout optimization (Faramarzi and Afshar 2014).

Sizing Optimization Sizing optimization problems are mathematical design optimization problems where the objective function is to find the optimum cross-sectional areas, i.e., the structural thickness of the truss structure's potential elements. The geometric feature of the truss, the design variable, is typically the structural thickness bounded to positive real numbers. A sizing optimization problem of a 5-member plane truss (see, e.g., Cheng and Guo 1997, Lemonge and Barbosa 2004,

Ohsaki 2016, Patnaik et al. 1998), i.e., a two-dimensional truss structure consisting of five structural members, is shown in Figure 3.2.

Shape Optimization Shape optimization problems are mathematical design optimization problems where the geometric feature of the truss, the design variable, is the position of the truss structure's nodal coordinates (Kaveh and Talatahari 2009b). The set of nodal points existing in the discretized design domain (see Section 4.4) is required to be constant. A shape optimization problem of a 10-member plane truss (see, e.g., Cheng and Guo 1997, Ringertz 1985) is shown in Figure 3.3.

initial design optimized design

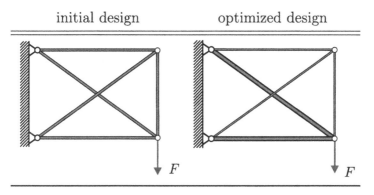

Figure 3.2 A structural sizing optimization problem of a 5-member plane truss: (left) an initial design; (right) an optimized design by manipulating the structural thickness bounded to positive non-negative real numbers. Note that the dimensions are not to scale. (Based on Tejani et al. 2018a)

initial design optimized design

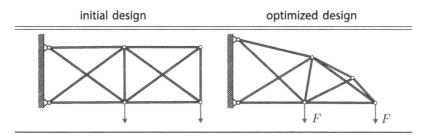

Figure 3.3 A structural shape optimization problem of a 10-member plane truss: (left) an initial design; (right) an optimized design by manipulating the truss structure's nodal coordinates and requiring the set of nodal points existing in the discretized design domain to be constant. Note that the dimensions are not to scale. (Based on Tejani et al. 2018a)

Topology Optimization Topology optimization problems are mathematical design optimization problems where the geometric feature of the truss, the design variable, is typically the structural thickness bounded to non-negative real numbers. In a discrete optimization like TTO it is possible to remove structural members and nodes from the initial design, which amounts to changing the topology of the truss by bounding the structural thickness to non-negative real numbers (Christensen and Klarbring 2008). In discretized TTO problems, the volume of the truss is minimized with an upper bound on the compliance, or the compliance is minimized with an upper bound on the volume (Stolpe 2010). A topology optimization problem of a 10-member plane truss is shown in Figure 3.4.

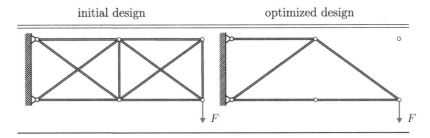

Figure 3.4 A structural topology optimization problem of a 10-member plane truss: (left) an initial design; (right) an optimized design by manipulating the structural thickness bounded to non-negative real numbers. Note that the dimensions are not to scale. (Based on Tejani et al. 2018a)

The advantage of using topology optimization instead of shape optimization lies in the fact that the set of nodes (connectivity) and therefore the topology of the truss are variable, so that possible unnecessary structural members and nodes can be removed from the initial design, i.e., the discretized design domain. This stimulates even more interest if the cost of the nodes is modeled (Kaveh and Zolghadr 2013). The coordinates of the nodal points are not considered as design variables; it is only assumed that the connectivity of nodes is variable, which can be interpreted as a possible change in topology; see Figure 3.4 right. Tejani et al. (2018a) concluded, on the one hand, that simultaneous consideration of sizing, shape, and topology optimization is the best approach to minimum weight problems. On the other hand, it comes at the expense of significantly increased computational complexity due to the use of binary, discrete, and continuous design variables. Despite this, the simultaneous consideration of all three kinds of truss optimization is reasonably assumed to be the best way to achieve a global optimum solution (Kaveh and Zolghadr 2013).

Complexity in the Engineering Design versus Computational Complexity

Complexity in the Engineering Design From the perspective of the iterative-intuitive engineering product design process, as proposed by Kirsch (1993b), topology optimization is the most complex and the most general problem of structural optimization because it considers all possible topologies of a given design domain rather than a prescribed or pre-processed topology and results in maximized material saving by selecting the optimal topology (Deb and Gulati 2001). Given the practical considerations of the iterative-intuitive engineering product design process, topology and shape optimization are managed independently, even though shape optimization is ideally a subclass of topology optimization from the viewpoint of mathematical design optimization. In contrast, topology and sizing optimization are dependent on each other, even though they are fundamentally distinct (Christensen and Klarbring 2008).

Computational Complexity From the perspective of computational complexity, truss design problems can be large-scale in terms of the number of constraints and variables, which is typical for the design of advanced truss structures (Stolpe 2016). For that reason, many theoretical and numerical challenges exist in this research field (Stolpe 2017). Yates et al. (1982) stated and proved that the problem of solving minimum weight truss design problems with discrete design variables is a member of the class of **NP**-hard problems. The simpler problem of minimum weight truss design problems subjected only to deflection is also proven to be **NP**-hard.

Symmetry in Structural Mechanics versus Symmetry in Mathematical Design Optimization
It is generally accepted in structural mechanics to use the concept of symmetry and antisymmetry to effectively optimize and analyze structural systems (Marsden and Ratiu 2013), see, e.g., Figures 5.4 and 7.8. In many real-world structural systems, symmetry and antisymmetry can be found because of long-standing established engineering practice, standardized manufacturing and assembly processes, and stability considerations.

Observation 3.1 (Symmetry in a Structural System).
The existence of symmetry in a structural system about an axis implies the existence of a design domain, boundary conditions, and load conditions that are symmetric around this axis.

Observation 3.2 (Kinematic Stability).
In general, apart from special deviations, solutions optimal for our MILP and QMIP problems (see Chapter 5) are kinematic indeterminate, not fully stressed[1], and include redundant structural members.

Remark 3.1.
We assume that stress and deformation of a(n) (anti)symmetric structural system is always (anti)symmetric.

Remark 3.2.
A truss including one or more redundant structural members or nodes cannot, in general, be fully stressed using a single material (Fuchs 2016).

Despite the engineering practice of using the concept of symmetry and antisymmetry, from the viewpoint of mathematical design optimization, symmetry is not a property of optimality such that optimal designs are excluded from the solution space if symmetry is enforced (Stolpe 2010). Explicitly, this means that the optimal solutions to the considered problems, in general, may not be symmetric even if the design domain, the external loads, and the boundary conditions are symmetric around an axis (Stolpe 2010). Achtziger and Stolpe (2007) concluded, very plausibly from the viewpoint of design optimization, that using engineering experience and intuition may not be expedient, since it excludes optimal designs from the solution space. Achtziger and Stolpe hypothesized using the results (optimal topologies) of a continuous relaxation as intuition. From the viewpoint of engineering, a severe limitation with this argument, however, is that symmetry in structural mechanics strongly depends on the design and the structure of the component, which shall be taken into account. In this thesis we assume, until further notice, that multiple, symmetric, or antisymmetric optimal designs of our MILPs and QMIP can exist even if variable linking of the possible symmetric structural members (see Section 5.4) is used in the problem formulation seeking symmetry around the horizontal or vertical mid-axis of the design domain.

---

[1] A (truss) structure where every present structural member is stressed to the minimum or maximum permissible stress of the material it is made of, i.e., the stress in each present structural member is at either the lower or the upper bound (Fuchs 2016, Stolpe 2017).

3.1.2 The Ground Structure Approach

According to Topping (1992), there are three kinds of mathematical programming methods for structural optimization at early design stages:

The Ground Structure Approach The thickness of the truss structure's potential members is bounded to non-negative real numbers, such that structural members can be excluded from the structure.

The Geometric Approach The coordinates of the nodal points and the structural member cross-sectional properties are considered as design variables.

Hybrid Methods Topological considerations are possible at specific points (areas) during optimization. Therefore, the design variables are assigned to two different groups or design domains. Hybrid methods could be applied to ground structures or use joint coordinates as design variables.

A truss structure consisting of a set of potential void or structural members and fixed nodal points is said to be the *ground structure*, whereas the approach that removes unnecessary structural members from the ground structure by optimization is said to be the *ground structure approach* (Ohsaki 2016). Assuming variable nodal points and structural members, intending to optimize the geometry of a truss, we obtain what is known as the *growing ground structure approach* (Hagishita and Ohsaki 2009). The ground structure defines the upper bound of the search space (Dorn 1964). Research has tended to focus on the ground structure approach rather than the growing ground structure approach (McKeown 1998, Tejani et al. 2018a).

Our TTO methodology, see Section 4.1, proceeds with the ground structure approach indicated in Dorn (1964), Bendsøe and Sigmund (2013), and Ohsaki (2016). With the advancement of LPs, computers, and software technology in the 1950s and 1960s, Dorn (1964) published the first work on the ground structure approach. More early studies from the 1960s on optimization of truss geometry and the ground structure approach are found in, e.g., Dobbs and Felton (1969) and Pedersen (1972, 1973). A more recent review of the literature on this topic is given by the review articles Kirsch (1989a), Topping (1983) and the book by Rozvany (2014).

Figure 3.5 left shows a challenging 72-member truss problem that has been commonly investigated (see, e.g., Kaveh and Talatahari 2009c, Lemonge and Barbosa 2004, Sedaghati 2005, Sonmez 2011). Figure 3.5 right shows a 25-member transmission truss problem that has been solved for different loading scenarios (see, e.g., Kaveh and Bakhshpoori 2013, Kaveh and Rahami 2006, Kaveh and Talatahari

2010b, Patnaik et al. 1998, Rao 2019). The important point to note here is that the ground structures are illustrated without boundary conditions and external loads. We state two spatial and no planar ground structures for the treatment of the more general (spatial) case.

ground structure I	ground structure II
72-member spatial truss	25-member spatial truss

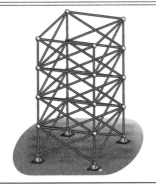 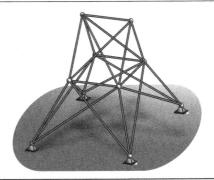

Figure 3.5 Two spatial ground structures [2]: (left) the ground structure of a 72-member spatial truss; (right) the ground structure of a 25-member spatial truss. Note that the dimensions are not to scale and the load conditions are left out. (Obtained from Camp and Bichon 2004, Kaveh and Talatahari 2009c)

Following Bendsøe and Sigmund (1995), the binary question of general topology and shape optimization of continuum structures is whether material exists for every point in the design domain. In studying the special case TTO, for simplicity we only consider potential void or structural members and not every point in the design domain; see Figure 3.6. This observation, when looked at from a more general point of view, leads to the ground structure approach, which allows the prediction of the layout of a truss structure existing of connections between a fixed set of frictionless nodal points as potential void or structural member, while the topology of the structure is not fixed a priori (Bendsøe and Sigmund 1995).

First, a design domain taking into account an external loading scenario and suitable boundary conditions is defined; see Figure 3.6 left. Second, the design

---

[2] It is not necessary to have exactly the same structural members as shown in the ground structure. Problem knowledge is used to exclude structural members from an original highly connected ground structure to produce the shown ground structure with reduced complexity.

domain is discretized with a set of frictionless nodal points distributed over the design domain; see Figure 3.6 middle. Afterward, the set of nodal points is connected with straight structural members constructing the ground structure. The ground structure is used to transmit concentrated or concurrent or both external forces, normal (active) single forces, and nodal forces inclusive of the bearing reaction force components. The center of forces coincides with the centroid of the volume of the nodal points, and the external forces have to be decomposed.

The solution for a TTO for time-dependent static loads is shown in Figure 3.6 right (Kuttich 2018). The ground structure applies to the famous example of the solar cells of the International Space Station (ISS), which are mounted on the main module using spatial trusses; see Figure 3.7.

We assume planar ground structures to form a subclass of spatial ground structures with the same intrinsic characterization. In case of a planar ground structure, constraints and variables corresponding to the missing spatial direction can be removed from our MILP and QMIP models (see Chapter 5) in a pre-processing step. To improve the efficiency of the ground structure approach, we separately implement a two and three-dimensional version of the ground structure approach in our models. All additional (linearized) constraints to implement geometry-based design rules for AM are considered in both ground structure approaches. This aspect will be dealt with in greater detail in Chapter 5. The decomposition of external forces is done in a pre-processing step for all our models. Our Ansys SpaceClaim add-in

design domain   ground structure   optimized design

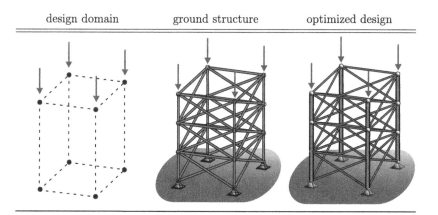

Figure 3.6 A TTO problem: (left) design domain of the idealized spatial truss ground structure mounted at the ISS (middle) ground structure with 48 potential structural members; (right) TTO solution. Note that the dimensions are not to scale. (Obtained from Kuttich 2018)

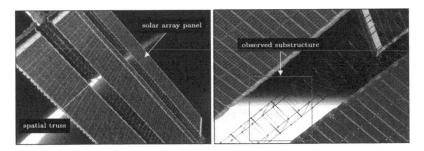

Figure 3.7 (Left) ISS solar array panels attached to the main module using spatial trusses, status expedition 17 2008 (Photo © ESA/NASA); (Right) Close-up view of a section of the ISS solar array panels, status expedition 40 2014 (Photo © ESA/NASA)

constructTOR in combination with Ansys SpaceClaim implements the ground structure (two and three-dimensional) using Algorithm 2 described in Chapter 6, which iterates over the set of nodal points creating work points representing the discretized design domain (see, e.g., SpaceClaim Corporation 2014, 2019, and the references therein). The set of potential void or structural members are implemented by volume bodies or beam objects, see Algorithms 3 and 4 described in Chapter 6. The 3D-CAD aspect will be dealt with in greater detail in Chapter 6.

3.1.3   Truss Topology Optimization

General Form of a Truss Topology Optimization Problem
The following model has been obtained from Tejani et al. (2018a). We define B_i to be a binary variable indicating whether the structural member i is a valid element of a truss. Let $X_i, \rho_i, L_i, A_i, E_i, \sigma_i, \sigma_i^{cr}$ denote the design variable, mass density, element length, cross-sectional area, Young's modulus, actual stress, and Euler's critical buckling stress of the structural member i, respectively. Let ξ_j, δ_j, b_j be the position, displacement, and mass values of node j, respectively. The structural natural frequency of the loading instance r is given by f_r.

As usual, the superscripts min, max, and $comp$ denote the minimum and maximum permissible stress as well as the compressive stress. The Euler buckling coefficient k_i is pre-processed using the cross-sectional area A_i of structural member i (Tejani et al. 2018a). With the above discussion, a simultaneous optimization of sizing (structural member cross-sectional area) and nodal coordinates (node position) against static loads is possible. We now state the generalized model of TTO, TTO_g, as a Nonlinear Programming (NLP)[3] based on the ground structure approach:

$$(\text{TTO}_g): \quad \min \sum_{i=1}^{m} B_i X_i \rho_i L_i + \sum_{j=1}^{n} b_j \tag{3.1a}$$

$$\text{s.t.} \quad |B_i \sigma_i| - \sigma_i^{max} \leq 0 \qquad\qquad \forall i = 1, \ldots, m \tag{3.1b}$$

$$|\delta_j - \delta_j^{max}| \leq 0 \qquad\qquad \forall j = 1, \ldots, n \tag{3.1c}$$

$$|B_i \sigma_i^{comp}| - \sigma_i^{cr} \leq 0 \qquad\qquad \forall i = 1, \ldots, m \tag{3.1d}$$

$$f_r - f_r^{min} \geq 0 \qquad\qquad \forall r = 1, \ldots, o \tag{3.1e}$$

$$A_i^{min} \leq A_i \leq A_i^{max} \qquad\qquad \forall i = 1, \ldots, m \tag{3.1f}$$

$$\xi_j^{min} \leq \xi_j \leq \xi_j^{max} \qquad\qquad \forall j = 1, \ldots, n \tag{3.1g}$$

$$\textit{check validity of the truss} \tag{3.1h}$$

$$\textit{check kinematic stability of the truss}. \tag{3.1i}$$

Furthermore,

$$\sigma_i^{cr} = \frac{k_i A_i E_i}{L_i^2} \tag{3.2}$$

applies. The objective function of the NLP TTO_g (3.1a) is to create a truss, which is minimal in terms of material. We call the model TTO_g a minimum weight truss design problem. In order to model the mass values of the nodal points, it is necessary to add a penalty function to the objective function consisting of the penalty parameter (mass value of a node) b_j. Prager (1974) introduced the cost (mass) of a nodal point to overcome the practically relevant problem of the existence of too many structural members.

Constraints (3.1b) to (3.1e) are often called the behavior constraints, whereby Constraint (3.1b) ensures that the actual stress σ_i of a structural member i is bounded to the maximum permissible stress σ_i^{max} of structural member i. Furthermore, Constraint (3.1c) ensures that the nodal displacement δ_j of a node j is bounded to the maximum permissible nodal displacement δ_j^{max} of a node j. Euler's critical load, also referred to as Euler buckling constraint (see Subsection 2.1.3), is described

---

[3] A NLP exists if the objective function is nonlinear or both objective function and feasible region are determined by nonlinear constraints.

by Constraint (3.1d). We emphasize that σ_i^{cr}, Euler's critical buckling stress of the structural member i, depends quadratically on the element length of structural member i, see Equation (3.2). To avoid destructive effects in the context of natural frequencies, e.g., free vibrations without damping, Constraint (3.1e) bounds the natural frequency f_r for multiple loading instances r (see, e.g., Achtziger and Kočvara (2008), Ohsaki et al. (1999) for time-independent models considering structural vibrations as free vibrations).

Constraints (3.1f) and (3.1g) are often called the side constraints, whereby (3.1f) merely specifies an upper and lower bound on the cross-sectional area A_i of a structural member i and (3.1g) specifies an upper and lower bound on the position ξ_j of a node j. The shape of the optimized structure depends significantly on the position of the nodes. Constraints (3.1h) and (3.1i) check the validity and kinematic stability (see Section 4.2) of the optimized structure, which have to be verified through numerical simulations (FEA) and experiments so that the truss does not generate a deformation mechanism; see Figure 3.8. If the truss has one or more deformation mechanisms (kinematic indeterminacy), we penalize the solution by assigning a large static penalty and generate an output. If the truss does not have a deformation mechanism, we carry out a more detailed numerical simulation of the truss, e.g., natural frequencies, element stresses, nodal displacements, and Euler's critical buckling stress of the truss. A flowchart of the generalized problem of TTO, TTO_g, can be seen in Figure 3.8.

Engineering Application

Lack of attention to manufacturing methods and to the requirements of structural design practice has led to actual engineering application being absent in the field of truss optimization (Stolpe 2016). In addition, the solutions of topology optimization are often unrealistic, e.g., not manufacturable, due to implied unstable nodes, unwanted mechanisms, intersection of structural members, and extremely slender structural members. Oberndorfer et al. (1996) addressed the problem that the implementation of strength, stability, and slenderness constraints in truss optimization models is indispensable to solve real-world engineering design problems. In addition, he showed that truss optimization including buckling constraints results in completely different optimal topologies by means of a bridge design and due to a stiffened membrane problem.

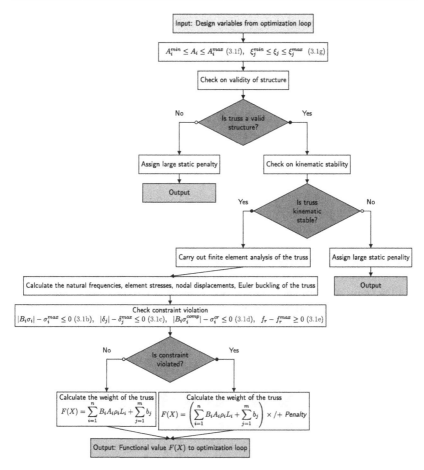

Figure 3.8 Flowchart of the generalized problem of TTO TTO_g. (Obtained from Tejani et al. 2018a)

Ohsaki (2016) stated that the ground structure approach is said to be the standard procedure for truss optimization, even though there are many difficulties in engineering application:

1. Too many structural members and nodes are needed in the initial ground structure, because only removal of structural members is possible, and addition of

structural members and nodes is very difficult without resorting to a heuristic approach.

2. The optimal topology depends greatly on the initial design, and an infinite number of nodes and structural members is needed if the nodal locations are also to be optimized.

3. Unrealistic optimal solutions with very long structural members, intersection, and overlap of structural members or a combination of these, etc., are often obtained.

4. The truss becomes unstable due to the existence of a node connected by two colinear structural members only, if too many structural members are removed.

Topologies classified as infeasible or unrealistic can be seen in Figure 3.9. We emphasize that this classification depends strictly on the manufacturing method used. In the case of powder-based additive (layer) manufacturing (see Section 2.2), an intersection of structural members is no manufacturing limitation due to the process of joining material layer after layer. Besides, the minimal and maximal manufacturable (self-supporting) structural members depend decisively on the used layer thickness, minimal and maximal (self-supporting) overhang of a component to the build platform, and the upskin and downskin angle of the component to the build platform; see Section 5.3.

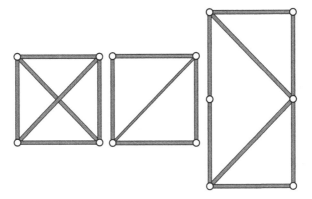

Figure 3.9 Topologies usually classified as infeasible or unrealistic: (left) intersection of structural members; (middle) very thin structural member; (right) unstable node. (Obtained from Ohsaki 2016)

3.2 Related Work

This section provides an overview of the related work in the context of mathematical programming methods for the optimization of truss structures. From an engineering perspective, a more comprehensive overview of truss optimization can be found in Tejani et al. (2018a). From a mathematical programming perspective, an overview is given by Arora et al. (1994), Stolpe (2016), Topping (1983). Besides, in Topping (1992), the author focuses exclusively on mathematical programming techniques. Since global optimization methods for the optimal design of truss structures make extensive use of the mathematical properties of the given problem, extending the methods to other cognate truss optimization problems is a challenging task (Stolpe 2016). Intuitively, from a mathematical programming perspective, this complexity appears natural, whereby from an engineering perspective, the task of optimizing different component variants (trusses) is established and minor. Consequently, from an engineering perspective, the global optimization methods are limited to special applications, and therefore a precise classification based on application cases (structural optimization variables) instead of types of mathematical programs is vital. To show both perspectives, we first review the existing literature regarding global optimization methods for truss optimization. Afterwards, we discuss the engineering perspective on the basis of the three kinds of structural optimization: sizing, shape, and topology optimization. This review does not claim to be a comprehensive summary of the truss optimization literature. For a general and comprehensive review of truss optimization, please refer to Bendsøe (2009) and the references therein. For an overview of global optimization methods for truss optimization, we recommend Stolpe (2016).

3.2.1 Literature Review

The first article known to the author that ever appeared about structural optimization is Galileo Galilei's study on the shape optimization of a cantilever in his famous 1638 publication *Discourses on the Two New Sciences* (Galilei 1914). While being interested in finding a mathematical theory of material, Galileo Galilei observed that part of the material of a cantilever could be removed without affecting the structural member's strength. Considering the breaking strength of a cantilever and a single concentrated load at the cantilever's upper tip, the result of the shape optimization conducted by Galileo Galilei corresponds to the shape of a square root function (Timoshenko 1983).

Over a century ago, Michell (1904) published key literature on topology optimization of least weight layouts of trusses, the so-called Michell structures, using
lower bounds on material consumption (costs) subjected to maximum stress and
displacement constraints (see, for instance, Hemp 1966, Rozvany 2009). Michell
used his derived method called optimality criteria. Hemp (1973) developed modified optimality criteria (see, e.g., Rozvany 2012, Rozvany and Gollub 1990) which
lead, in general, to lighter trusses. In addition, Hemp stated the major disadvantage
that the original optimality criteria stated by Michell are only valid for a highly
restricted class of support conditions. Another well-known criticism of Michell's
findings is that discretized Michell structures, with their infinity of possible structural members, have limited real-world application. Some comments on the practical
usefulness of Michell structures are discussed in Prager (1974, 1977, 1978), Rozvany (1996b) and the references therein. As others have highlighted, Habermehl
(2013), Mars (2014), Michell structures are still—discounting limited real-world
application—a benchmark (see, e.g., Sigmund et al. 2016) for the results of modern
computations applying numerical methods to challenging large-scale truss design
problems. In his seminal paper of 1964, Dorn (1964) introduced the ground structure
approach combined with a continuous design variable for the cross-sectional area
of a structural member and laid the foundation for TTO. Stolpe (2016) highlights
that Dorn presented the first single load minimum weight LP problem. For further
details on the LP problem, please refer to Hemp (1966). The first systematic study
on constrained structural optimization problems under multiple loading conditions,
having singular global minima, was carried out by Sved and Ginos (1968). They
showed that excluding structural members from the ground structure is necessary to
guarantee the possibility of finding the global optimum. Topping (1983) proposed a
review of mathematical programming methods used in sizing, shape, and topology
optimization of truss structures. He concludes that algorithms with the possibility
of introducing new structural members and nodes to the ground structure (see also
Subsection 3.1.2) during the optimization procedure have yet to be developed.

Ringertz (1985) suggests a two-stage optimization of trusses subjected to a single
loading case, displacement, and stress constraints. He used LP for static analysis
and, subsequently, NLP for further, more detailed optimization. Besides, Ringertz
(1986) presented a nonlinear branch-and-bound algorithm for TTO subjected to
displacement and stress constraints. In his article on a genetic algorithm for TTO
considering nodal costs, Ohsaki (1995) concluded that nodal costs significantly
impact real-world engineering design problems[4]. Nevertheless, the proposed algo-

---

[4] We remark that this is the case for traditional manufacturing methods but not (in general)
for AM.

rithm is stated to be computationally expensive, which excludes practical relevance. A procedure for a combined sizing, shape, and topology optimization of spatial trusses was first evolved by Rajan (1995).

The first attempt to employ a simulated annealing based approach for structural optimization subjected to multiple loading conditions, buckling, and displacement constraints or both was investigated by Dhingra and Bennage (1995). Zhou (1996) proposed a discussion on topology optimization subjected to local buckling constraints and the ground structure approach. Unlike other research (see, e.g., Gengdong and Zheng 1994) carried out in this research area, Zhou concluded that adding system buckling constraints via a geometrical optimization of the truss structure's nodal coordinates may result in difficulties, e.g., error-prone solutions which cannot be avoided without manipulating the ground structure (Rozvany 1996a). Deb and Gulati (2001) provided a genetic algorithm-based optimization procedure for simultaneous size, shape, and topology optimization of plane and spatial trusses. Finotto et al. (2013) used nonlinear FEA, a genetic algorithm, and fuzzy logic to state a method for discrete topology and sizing optimization of trusses including cables.

3.2.2   The Global Optimization Perspective

In recent years, there has been considerable interest in solving global optimization problems for structural (truss) optimization and engineering design and control (Arora et al. 1994). Stolpe (2016), an authority on mathematical truss optimization, outlines the primary reason for this as being that sizing or topology optimization of truss structures or both is ideal to combine structural design optimization problems with other research areas such as mathematical programming, operations research, and discrete optimization. Agreeing entirely with Stolpe, and considering the highly likely computational complexity of truss optimization problems, any assistance from the aforementioned research areas would be welcome in the structural optimization field. A quote from Yates et al. (1982) reflects the high computational complexity of truss optimization, especially the computational complexity results of the classical DMT[5] and DMTD[6] problem:

---

[5] Discrete structural member-size minimum weight truss problem.

[6] Discrete structural member-size minimum weight truss problem with deflection constraints.

The problem of minimizing the weight of structural trusses subject to constraints on joint deflection and structural member stress and gauge has been considered. In particular, the case DMT where structural member sizes are available from a discrete set has been examined and shown to be NP-hard. As an intermediate step, the problem DMTD, where only joint deflection constraints apply, was also shown to be of equivalent difficulty. Consequent upon these results, the problem of devising an approximation algorithm for DMT that guaranteed a bound on the absolute error was considered, but unfortunately, this too proved to be NP-hard.

Analyzed from the perspective of mechanical engineering, intuitively, this theoretical complexity appears natural, considering the related problems, e.g., flexible structural members in multibody dynamics or a constitutive model of hyperelastic material, which are generally considered as challenging (not necessarily theoretically complex) engineering problems implying a high degree of nonlinearity in the physics of the problems. Since sizing or topology optimization or both of truss structures are intrinsically non-convex, many global optimization methods (see Achtziger and Stolpe 2007, Floudas and Pardalos 2013, Wolsey and Nemhauser 1999 and references therein) rely on equivalent problem (re-)formulations (Stolpe 2016). Stolpe points out that the most used global optimization method for optimization of truss structures is the branch-and-bound algorithm (Lawler and Wood 1966). He also states that for most heuristics, the solution quality is not validated sufficiently, such that it is necessary to use numerical simulations (Holzapfel 2000, Wriggers 2008) to analyze, characterize, and predict the mechanical behavior of the optimal truss structure. Deterministic approaches for global optimization and a set of interdisciplinary applications are given in Horst and Pardalos (2013), Horst and Tuy (2013). Other noteworthy solution approaches for the optimization of truss structures are genetic algorithms (Hajela and Lin 1992, Rajeev and Krishnamoorthy 1992, Toğan and Daloğlu 2006), particle swarm optimization methods (Kaveh and Talatahari 2009b, Li et al. 2009), harmony search algorithms (Lee et al. 2005), ant colony optimization (Bland 2001, Kaveh et al. 2008), artificial bee colony algorithms (Sonmez 2011, Stolpe 2011), big bang-big crunch optimization (Camp 2007, Kaveh and Talatahari 2010a), tabu search (Bennage and Dhingra 1995a), and simulated annealing (Bennage and Dhingra 1995b, Kripka 2004). In the following, we give a brief overview of equivalent problem (re-)formulations for the non-convex optimization of truss structures.

Linear and Quadratic Programs

A broad class of truss optimization problems can be re-formulated as a MILP problem or Mixed-Integer Quadratic Programming (MIQP) problem. Nowadays, a key research gap (see, e.g., Stolpe 2016, Tejani et al. 2018a, for other research gaps) is to use available computationally efficient branch-and-cut solvers like CPLEX and GUROBI to solve large-scale linear truss optimization problems under consideration of structural design practice. The purpose is to generate actual real-world engineering applications, e.g., by implementing design rules for AM. A disadvantage regarding the usage of LP or MILP programs is that, in the case of multiple loads, the optimal truss structure is generally neither statically determinate nor fully stressed.

Toakley (1968) stated an optimization of stiff-jointed frameworks using rigid-plastic theory and an optimization of statically determinate triangulated frameworks subjected to deflection constraints using discrete programming techniques. As mentioned by Burns (2002), Toakley perceived the need for discrete structural optimization and applied the branch-and-bound method, Gomory's cutting plane method (Gomory 1960, 1963), and new heuristic methods to truss optimization problems.

Reinschmidt (1971) introduced the basic idea of using the branch-and-bound method to solve a linearized re-formulation of the NLP problem of plastic design of building frames. Bauer et al. (1981) state a cost (weight) minimization of spatial trusses subjected to a single displacement constraint. Discrete structural member cross-sectional areas are used. Applying the Galerkin procedure (Reddy 2017) and using binary variables, the final formulation is a $0-1$ LP problem. The problem was solved using Balas's Additive Algorithm (Balas 1965, Glover and Zionts 1965) for IP problems with binary variables. Another way of stating truss optimization problems is the re-formulation as a MINLP problem, see, e.g., Stolpe (2007, 2016) for an overview of re-formulations applied towards the design of continuum structures and composite materials. By using the so-called re-formulation-linearization technique, Faustino et al. (2006) stated a $0-1$ ILP problem to find kinematically stable truss structures with optimal topology and minimum volume. The numerical results provided by Faustino et al. strongly suggest that a branch-and-bound algorithm was computationally too demanding for large-scale structural optimization. In his seminal article on the re-formulation of topology optimization problems as linear or quadratic mixed $0-1$ programs, Stolpe (2007) defines a maximum stiffness TTO problem. Three benchmark problems with up to 200 structural members in a plane ground structure (Stolpe 2007) are used to measure the performance of CPLEX (2014). The computational results identified that a conventional branch-and-cut method implemented in CPLEX (2014) can solve medium sized real-world maximum stiffness TTO problems.

Kanno and Guo (2010) showed that a robust TTO problem subjected to stress constraints, uncertain loads, and discrete structural member cross-sectional areas can be reduced to a MILP problem. The main limitation of the MIP problem of Kanno and Guo is that it is inapplicable for large-scale structural optimization, since the MIP

problem includes large numbers of variables and constraint conditions. Thus, the
results obtained by the robust TTO MILP problem are merely capable of being used
as benchmark problems to validate other problem (in)dependent (meta-)heuristics,
optimization algorithms, and methods. The proposition of new benchmark problems
is a known research gap in TTO (Gandomi and Yang 2011, Tejani et al. 2018a). It
has been emphasized by Kanno and Guo (2010) that the global optimal solution for
the robust TTO can be computed using a conventional branch-and-cut solver, even
if so far not practically relevant.

Ehara and Kanno (2010) investigated a numerical method for optimizing tensegrity[7] structures (Fuller 1982) based on the ground structure approach and two
sequentially used MILP problems. First, the number of struts subjected to the self-
equilibrium conditions and the discontinuity conditions of the struts are maximized;
secondly, the number of cables is minimized. Topology optimization and the development of innovative tensegrity structures in real-world applications via MIP are
presented in Kanno (2012, 2013a, b). It has been demonstrated that an LP problem
formulation of the minimum volume problem only subjected to stress constraints
can be used for the initial design of truss structures (Stolpe 2017).

Conic Programs
Several types of truss optimization problems, e.g., the minimum compliance problem, can be reformulated as a SemiDefinite Programming (SDP) problem. SDP
problems are a special case of Conic Programming (CP) problems and are solv-
able by interior-point methods. Nowadays, the key advantage of CPs is that, unlike
LPs, in-depth real-world design problems, such as natural (structural) frequencies
or local and global buckling can be reformulated as SDP programs (Stolpe 2017).
The major disadvantage is that, even though since the 1990s most interior-point
methods for LP problems have been well adapted for SDP and both methods have
polynomial worst-case runtime, i.e., are in the complexity class **P**, solvers for SDP
problems are still in the early stages of development (Stolpe 2016). Initial work in
this field focused primarily on minimizing the maximum compliance subjected to
equilibrium constraints and restrictions on the component volume (see, e.g., Ben-Tal
and Nemirovski 1994, 2000, Vandenberghe and Boyd 1996).

Ben-Tal and Nemirovski (1997) introduced and motivated robust Truss Topology
Design (TTD) (see, e.g., Achtziger et al. 1992, Ben-Tal and Bendsøe 1993, Ben-
Tal and Nemirovski 1994, Ben-Tal et al. 1993, Bendsøe et al. 1994 and references

---

[7] Pin-jointed structure consisting of continuously connected tensile structural members
(cables) and disjointed compressive structural members (struts) subjected to prestress at the
initial (stable) equilibrium state (Kanno 2012).

therein) via SDP to improve applicability in engineering. A truss is called robust[8] (Mulvey et al. 1995) if it is rigid while being subjected to two kinds (finite sets) of loading scenarios; the first finite set defines a static multi-load problem, the second finite set defines small static and uncertain (amount and position) loads which may apply on any node.

In Ohsaki et al. (1999), the optimum design problem of trusses subjected to the eigenvalue of vibrations (see also Khot 1985) is implemented via an SDP problem. In addition, Ohsaki et al. presented an algorithm specialized in optimum design with multiple eigenvalues. Achtziger and Kočvara (2008) stated a minimum volume (weight) problem, minimum compliance problem, and the problem of maximizing the minimal eigenvalue of the truss as SDP problem (see also Lewis and Overton 1996, Seyranian and Mailybaev 2003 for practical applications). Besides, multiple mass conditions are implemented, which encourages the practical usefulness of the SDP problems. Cerveira et al. (2009) presented two special-purpose branch-and-bound based algorithms and solved the maximum stiffness problem as Mixed-Integer SemiDefinite Programming (MISDP) to global optimality for a small planar ground structure with a maximum of nine nodes and 20 potential structural members.

Kočvara (2010) showed that the minimum volume (weight) TTO problem subjected to stiffness and vibration constraints can be formulated as fully equivalent to a Integer SemiDefinite Programming (ISDP) problem rather than as a MINLP problem. The performance was rather disappointing. Kočvara argues that this was probably as a result of underdeveloped solvers (see Tawarmalani and Sahinidis 2005), and emphasized that Integer Conic Programming (ICP) needs more attention from the research area of structural optimization. Since truss optimization with continuous design variables is unusable for manufacturability reasons using classical manufacturing techniques, Kanno (2016) proposed a minimum compliance TTD problem subjected to the structural volume (weight) and a limited number of different cross-sectional areas as a Mixed-Integer Second Order Cone Programming (MISOCP) problem. Kanno provides a large set of benchmark problems for structural optimization using planar ground structures with up to 748 potential structural members, which are solved using a branch-and-cut algorithm adapted for MISOCP.

An extension of the SDP problem of Ben-Tal and Nemirovski (1997) to discrete structural member cross-sectional areas, vibrations, and positioning of active components[9] via a MISDP problem was done by Kočvara (2010) and Mars (2014). Habermehl (2013) applied the concept of robust TTD subjected to active com-

---

[8] Note that we introduce a fundamentally different interpretation of robustness in Section 5.5.

[9] Usage of piezo-ceramic staple actuators and hydro-pneumatic spring-damper systems to allow the truss to actively react on external (uncertain) effects.

ponents. Besides, Gally et al. (2015) proposed further extensions to the choice of connection element types (hinges or rigid connections) for beam structures. In Gally et al. (2018), an approach to optimally place active structural members (see also Hiramoto et al. 2000, Schäfer 2015) for buckling control subjected to local buckling constraints is presented. Kuttich (2018) developed a robust topology- and feedback-controller design for linear time-invariant systems via SDP.

Nonlinear Programs
The last alternative is to reformulate a truss optimization problem as a MINLP problem. Since the continuous relaxation is in general non-convex, the MINLP is apparently hard to solve to global optimality (Stolpe 2016). Pyrz (1990) introduced a discrete optimization of elastic shallow spatial trusses using geometric nonlinear modeling. The minimum volume (weight) truss optimization problem was subjected to structural member cross-sections, element stresses, element stability, and global structural stability. Pyrz used the enumeration method of Greenberg (1971) to solve the problem, concluding that the effectiveness of the method is limited by the number of design variables and the number of available cross-sectional areas, which is inadequate for optimization of real-world engineering design problems. The largest problem stated by Pyrz, an elastic shallow dome structure, has seven available cross-sectional areas and four design variables.

Stolpe and Kawamoto (2005) stated a non-convex MINLP problem for a simultaneous analysis and synthesis (design) of articulated mechanisms, modeled as truss structures, subjected to large displacements. A special purpose branch-and-bound method for solving the non-convex MINLP problem and convex NLP relaxations are presented. The largest benchmark problem is a simultaneous topology and geometry optimization of a force inverter[10] modeled as a planar ground structure with 25 nodes and 300 potential structural members. Stolpe and Kawamoto suggest that the non-convex MIP problem and their special purpose branch-and-bound method can be used to reliably solve mechanism design problems of realistic size to global optimality.

In Achtziger and Stolpe (2007), ten single-load and four multiple-load (e.g., bridge, Michell member, cantilever member) benchmark problems are presented (see Achtziger and Stolpe 2008, 2009, for the theory of the relaxed problems and implementation details). The discrete minimum compliance problem for a single load Michell member using a ground structure consisting of 40 nodes and 632 possible structural members is solved without problem size reductions. Achtziger

---

[10] Mechanism for which the applied external force at a node is balanced out by the displacement of the output node.

and Stolpe used variable linking for the possible symmetric structural members in the problem formulation to request symmetry around the horizontal mid-axis of the design domain and compared the optimal structures subjected to and without symmetry. For information on our design-variable linking technique, we refer to Section 5.4, and Figures 5.4 and 7.8.

3.2.3   The Engineering Perspective

Sizing Optimization

Rajeev and Krishnamoorthy (1997) presented genetic algorithm-based methodologies for size or topology optimization or both. In the case of size optimization, they analyzed a 10-member planar truss problem under two point loads with constant configuration (position of joints) and an 18-member planar truss problem under five point loads with variable configuration. Tang et al. (2005) used a genetic algorithm for design optimization of truss structures, considering sizing, shape, and topology variables to optimize, inter alia, the weight of a 15-member planar truss under a single tip load subjected to stress constraints.

Fenton et al. (2014) argue that most methods for truss optimization lack practical usability as they only focus on minimizing cross-sectional areas, neglecting crucial section properties, material specifications, design codes, standards of practice, and manufacturing technologies. For this reason, the authors investigated a genetic programming (especially grammatical evolution) based truss optimization approach to optimize a 10-member and 17-member planar cantilevered truss problem. Prior to Fenton et al., both truss optimization problems were investigated in Kaveh and Talatahari (2009a), Lee and Geem (2004), Li et al. (2007), Luh and Lin (2011).

To gain practical, real-world relevance, Gonçalves et al. (2015) presented a search group algorithm for truss optimization problems subjected to natural frequencies, stress, and local buckling constraints. To show the computational performance, six truss optimization benchmark problems subjected to multiple cases are stated (see Gomes 2011, Kaveh and Javadi 2014, Kaveh and Zolghadr 2014, Lingyun et al. 2005, Miguel and Miguel 2012, Wang et al. 2004, Wei et al. 2011, Xu et al. 2003, for different cases). It has been suggested by Gonçalves et al. (2015) that further development and application to real engineering problems, especially the optimization of planar and spatial, sway and non-sway frame structures (Hasançebi and Azad 2012, Kaveh and Nasrollahi 2014, Maheri and Narimani 2014) is necessary.

De Souza et al. (2016) stated a methodology for size and topology optimization of transmission line towers. To ensure industrial application, different pre-established topologies based on the design practice and feasibility of prototype testing are used.

De Souza et al. benchmarked size optimization with size and topology optimization and size and shape optimization with topology optimization for two different applications. The first one is a classical transmission line tower, whereas the latter is a 115 kV transmission line tower. As a real-world case of loading, a cable conductor rupture scenario and a wind load hypothesis are used. In addition, the standard design codes of ASCE/SEI 10–15 (2015) are applied. Mortazavi and Toğan (2016) stated a sizing optimization of a 582-member spatial tower considering the symmetry of the truss structure. The standard design codes of the American Institute of Steel Construction (1989) are implemented.

Sizing and Shape Optimization
Rajeev and Krishnamoorthy (1997) analyzed an 18-member planar cantilever truss subjected to stress, displacement, and Euler buckling constraints. Hasançebi and Erbatur (2001) optimized a 47-member planar truss tower (see also Hansen and Vanderplaats 1990, Salajegheh and Vanderplaats 1993) subjected to three different loading scenarios. To increase the computational efficiency of the optimization, they introduced two new methodologies: annealing perturbation and adaptive reduction of the design space using an improved genetic algorithm. To decrease the number of independent size variables, the symmetry of the truss is used.

Due to manufacturing limitations and standardization, cross-sectional areas are usually defined by discrete values in practice. To fulfill this condition and gain practical relevance, Tang et al. (2005) and Miguel et al. (2013) stated a mixed variable truss optimization problem considering continuous nodal coordinates and a discrete set of cross-sectional areas. The mixed variable truss optimization problems are solved by a genetic algorithm or firefly algorithm, developed by Yang (2009). A 15-member planar truss and a 25-member spatial truss are optimized. Rahami et al. (2008) analyzed an 18-member planar truss, whereas Gonçalves et al. (2015) optimized a 37-member planar truss with natural frequency constraints. Ahrari and Deb (2016) optimized a 77-member planar bridge truss adapted from (Hasançebi 2008).

Sizing and Topology Optimization
Sved and Ginos (1968) formulated a NLP problem and optimized a 3-member planar pin joined structure subjected to three loading scenarios and concluded that excluding structural members from the ground structure is necessary to ensure the possibility of finding the global minimum.

Ringertz (1985) used LP to obtain a statically determinate basic structure and subsequently NLP for further minimization of the structural weight. Besides, he argues that one of the major drawbacks of previous findings (see Dobbs and Felton

1969) is that possible structural members which approach zero were removed from the structure after optimization using heuristic methods. Following Ringertz, on the one hand, this excludes the possibility of letting structural members rejoin the optimization process; on the other hand, the force distribution of the optimized truss differs from the real force distribution, which makes the optimized truss useless for real-world engineering design problems without numerical or experimental verification. He stated an optimization of a 10-member planar truss considering 6 nodal points, a 378-member truss considering 28 nodal points, and a 2415-member truss considering 70 nodal points. Nakamura and Ohsaki (1992) stated an optimization of plane trusses subjected to multiple fundamental frequencies. They focused on the avoidance of unrealistic slender structural members and argued that engineers demand realistic (designed for manufacture) trusses instead of (theoretical) optimal but impractical trusses. An optimization of a 36-node plane square truss, 55-node plane rectangular grid, and a 55-node plane are presented. Xu et al. (2003) argue that research has tended to focus on the costs of possible structural members rather than on the costs of nodes connecting the structural members, even though it is well known in engineering practice that the cost of the nodes may be equivalent or greater than the costs of the structural members[11]. Based on that experience, Xu et al. developed an algorithm-based TTO with structural member and nodal cost incorporated in the objective function. Kaveh and Zolghadr (2013) presented a sizing and topology optimization of trusses subjected to static and dynamic constraints. An optimization of a 24-member planar, 20-member planar, and 72-member spatial truss is presented. In Mela (2014), buckling constraints for steel trusses, according to Da Silva et al. (2012), are implemented via a MILP problem. A noteworthy aspect is that besides the Euler buckling formulation, more complex buckling constraints derived from design codes out of Da Silva et al. (2012) can be incorporated into the MILP problem.

Sizing, Shape, and Topology Optimization
Rajan (1995) was among the first who stated a methodology for simultaneous consideration of sizing, shape, and topology optimization of truss structures. Noilublao and Bureerat (2011) argued that an evolutionary algorithm-based multi-objective design problem of trusses would be valuable for the engineering research field and stated a simultaneous topology, shape, and sizing optimization of a three-dimensional slender truss tower. The first investigations into the growing ground structure approach for shape and topology optimization of optimal pin-jointed frames were carried out

---

[11] We remark that this is the case for traditional manufacturing methods but not (in general) for AM.

by McKeown (1998). He underlines that his method has the advantage that it can generate simpler, more realistic structures. A planar and symmetric cantilever truss and a planar cantilever truss with restricted shape are presented.

Bojczuk and Rębosz-Kurdek (2012) proposed a structural member exchange method for simultaneous optimization of truss shape and topology. To reduce computational effort, they focused on the optimization of equivalent sub-topologies, which apply for many real-world trusses, in particular symmetric trusses and trusses containing repeated structural member subsystems. Deb and Gulati (2001) introduced a methodology, which differentiates between basic (load-carrying) and non-basic (not load-carrying and not support) nodes, knowing from engineering design practice that basic nodes are indispensable for a feasible design and non-basic nodes are added for optional load sharing. They suggest that this methodology would lead to more manufacturable truss designs and would reduce the computational complexity. Deb and Gulati showed a 15-member planar, 11-member planar, 45-member planar, 39-member planar two-tier, and a 25-member spatial truss using their genetic algorithm for the design of minimum weight truss structures. In Tejani et al. (2018b), a simultaneous size, shape, and topology optimization of planar and spatial trusses is stated.

Bridging Algorithm-Driven Truss Optimization and Additive Manufacturing

4

Chapter 4 is divided into five subsections. In Section 4.1, our algorithm-driven TTO methodology for AM is presented. We extend the established FEA driven product design process by a preliminary design defined by a MILP or QMIP. Section 4.2 defines the basic problem statements in TTO and associated kinematic indeterminacy. Preparations towards a truss-like structure design problem are introduced in Section 4.3. The assumptions of our models regarding the displacements, the equilibrium of forces, and the moment equilibrium are introduced. A concept of design space constraints for an AM system is proposed in Section 4.4. Our assumptions regarding the assembly space of an AM system and our ground structure approach adapted for AM are introduced. In Section 4.5, our algorithm-driven optimization for a support-free component is compared with other methods in this field of research.

4.1 Algorithm-Driven Design Methodology for Additive Manufacturing

4.1.1 A Technical Operations Research Approach

To systematically explore the area of mathematical optimization (modeling), CAD (construction), FEA (validation), and the area of AM processes (realization), we adapt the methodology of Altherr et al. (2016). Figure 4.1 describes the development of an efficient, additively manufactured truss-like structure. The methodology is divided into two phases: the decision phase (1–3) and the action phase (4–7).

In the first step (1), customer and designer have to agree on a technical specification, which is a fair balance of loading case, costs, and manufacturability, inspired by Pelz et al. (2012). The loading case is determined by the practical use of the planned component and the associated component strength and thus taken to be immutable.

© The Author(s), under exclusive license to Springer Fachmedien Wiesbaden Gmbh, part of Springer Nature 2022
C. Reintjes, *Algorithm-Driven Truss Topology Optimization for Additive Manufacturing*, https://doi.org/10.1007/978-3-658-36211-9_4

The manufacturability depends on the chosen AM process and its standards. The objective function (2) minimizes the material, volume, and costs. An upper bound for the costs is determined by the selected ground structure, the chosen AM process, AM standards, and AM constraints (3). Furthermore, pre-processed structural member types with different cross-sections and costs, which are proportional to the volume and the material consumed, affect the upper bound. In the next step, a MILP or QMIP is used as optimization method to determine the optimal topology, i.e., the global optimum (4). The mathematical optimization results are implemented in our Ansys SpaceClaim add-in construcTOR and afterwards numerically analyzed via an FEA (5–6). The AM process set in step (3) is used for realization (7).

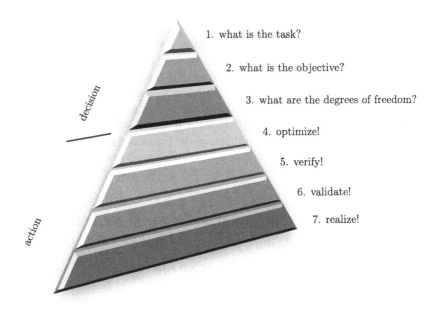

1. what is the task?

2. what is the objective?

3. what are the degrees of freedom?

4. optimize!

5. verify!

6. validate!

7. realize!

Figure 4.1 The TOR pyramid—a step by step guide towards an algorithm-driven product design process, based on Pelz et al. (2012)

The manufacturability and mechanical compliance with standards are ensured, since in step (3) all manufacturing restrictions are implemented as constraints in the MILP or QMIP. In order to increase the manufacturing accuracy of the chosen AM process, at the expense of costs, standard design rules for component manufacturing aiming at a high manufacturing accuracy (see, e.g., VDI 3405-3-3) can be included in the mathematical optimization models. These standard design rules will limit the solution space and thus counteract the "complexity for free" phenomenon of AM (Conner et al. 2014). In AM, complexity is said to be free, as the component is made layer-by-layer so that costs and manufacturing time are independent of the component complexity (Atzeni and Salmi 2012). To take full advantage of the phenomenon "complexity for free", standards should be used in a targeted manner, as we aim to overcome the creative human process of design and construction, with the help of a mathematically and algorithmically driven product design process, inspired by Pelz et al. (2012). In Chapter 7, we present four design studies based on the TOR approach.

4.1.2 An Algorithm-Driven Product Design Process

The established application of using the FEA as a design tool, see Figure 4.2, is to change the design process from iterative cycles of "design, prototype, test" (traditional product design process) into a streamlined process (FEA driven product design process) where prototypes are used only for final design verification. Through the use of FEA, design iterations are moved from the physical space of prototyping and testing into the virtual space of computer based simulations (Kurowski 2006). The goal is to establish a safe and cost-effective product design process.

We extend this FEA driven product design process by a preliminary design determined by our algorithm-driven product design process; see Figure 4.2 right. Mathematical optimization and FEA (simulation) is used concurrently within the design process, whereby mathematical optimization is used first for topology and shape optimization and FEA only for validation. The design and manufacturing experts using the algorithm-driven product design process decide which circum-

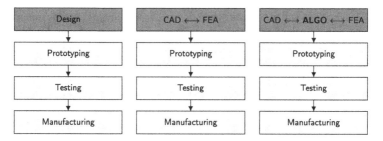

Figure 4.2 Traditional, FEA, and algorithm-driven product design process

stances are implemented by our TTO models (see Chapter 5), are pre-processed in our Ansys SpaceClaim add-in construcTOR (see Chapter 6), or are omitted and validated with an FEA. A data interface between the CAD neutral software Ansys SpaceClaim and a Computer-Aided Engineering (CAE) software enables a data conversion from a geometrical representation of a component (CAD file) into a triangulated surface of a component (Standard Triangle Language (STL) file). Once the STL file has been generated, the file is imported into a slicer software. This so-called slicer converts the STL file into a G-code file used for Computer-Aided Manufacturing (CAM).

The Ansys SpaceClaim add-in construcTOR serves as an interface between the TTO models, the FEA, CAE, and finally AM. Partial problems that cannot be modeled in the TTO models, depending on their natural constraints, have to be post-processed in the Ansys SpaceClaim add-in construcTOR. Considering our (quantified) MILP models, the post-processing of intersections and interferences for an FEA, see Chapter 6, are implemented by the Ansys SpaceClaim add-in construcTOR. Furthermore, the 3D-CAD data is prepared for an FEA inclusive geometry cleanup and simplification using the Ansys workbench. Finally, a verification of the component via a linear elastic and nonlinear elastic numerical FEA is possible. All necessary 3D-CAD data is available to slice and subsequently manufacture the component using any desired AM process. As the CAD neutral Ansys SpaceClaim add-in construcTOR completely processes the data and is integrated into the Ansys workbench, the design and manufacturing experts have access to the full range of standard CAD software via a neutral data exchange format like Sandard for The Exchange of Product model data (STEP) (ISO 10303-242:2020-04).

As shown in Figure 4.3, the process is iterative. The AM process and related standards (see lower left) influence the constraints of the selected TTO models and vice versa. The design and manufacturing experts are able to control the results of the optimization (SOL file) through the assembly (STEP file) automatically generated by the Ansys SpaceClaim add-in construcTOR (3D CAD tool). Subsequently, an FEA (validation) and numerical shape optimization of high-stressed connection nodes can be performed. Either the design results are not sufficient, so that the constraints of the TTO have to be adapted, or the component can directly be manufactured with AM. The processing step TTO encompasses all optimization methods available or planned within our algorithm-driven product design process. Using our (quantified) MILP models, instance variables such as the loading case, material specifications, and the reference volume are specified. The aim of the algorithm-driven product design process is to use our optimization at the first chronological level of the design process and FEA only for validation.

Figure 4.3 Schematic of our algorithm-driven product design process

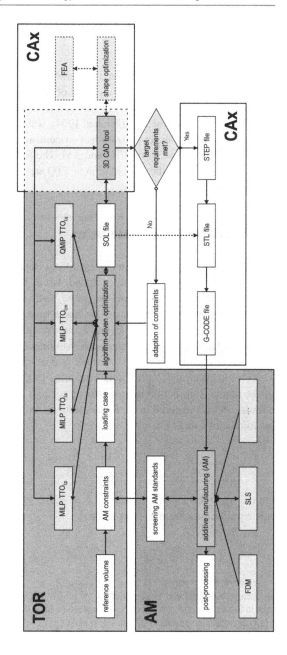

4.2   Basic Problem Statements and Kinematic Indeterminacy

There have been several studies on TTO, since the beginning of the 20th century. Since then, TTO has been a growing field of research, combining different areas of engineering with mathematics. In addition to classical applications such as the optimality of Michell structures (Michell 1904), which are used in civil engineering for the design and volume optimization of space structures, there have recently been remarkable efforts to combine TTO and AM. The following is a brief overview of the models that support the combination of TTO and AM, considering that we want to formulate (quantified) MILPs.

4.2.1   Basic Problem Statements

The following models have been obtained from Bendsøe (2009), whereas mechanisms and rigid body motion are excluded. Let E and V denote the set of edges (possible structural members) and the set of vertices (possible connection nodes) of a truss, respectively. We will denote that a truss $T = \{b_1, \ldots, b_m\}$ is a set of structural members b with $T \subseteq E$, i.e., T indicates at which edges in E a structural member is placed. Let A_b, L_b denote the cross-sectional area and length of a structural member b, respectively. It is assumed that all structural members are made of linear elastic materials[1], with Young's modulus E_b (Dym and Shames 1973) defining the relationship between stress (force per unit area) and strain (proportional deformation) in a material in the linear elasticity regime of a uniaxial deformation. The volume of a truss T is

$$V_T = \sum_{b \in T} v_b \,, \tag{4.1}$$

with the structural member volumes $v_b = A_b L_b$, $b \in \{b_1, \ldots, b_m\}$. As usual, the static equilibrium is expressed as

$$Bq = p \,, \tag{4.2}$$

---

[1] A linear elastic material is identified by its elastic potential, whereby only the quadratic terms in the strain persist. A linear elastic material can be defined in an isotropic, orthotropic, or fully anisotropic version. Isotropic linear elastic materials are identified by their Poisson's coefficient and Young's modulus.

where q is the structural member force vector, p is the nodal force vector of the free degrees of freedom, and B the compatibility matrix, which relate element displacements to system displacements (de Borst et al. 2012). The basic problem statement in terms of structural member forces (single load truss problem) is formulated as a stress based minimum mechanical compliance problem (TTO$_c$, see Table 4.1) using the minimum complementary energy principle (Bendsøe 2009). Problem TTO$_c$ is simultaneous convex in structural member forces and volumes.

The traditional formulation of TTO in terms of structural member forces (TTO$_{p;a}$, see Table 4.1), is valid for single load, plastic design (Bendsøe and Kikuchi 1993, Rozvany 1984, Topping 1993). This single load, plastic design problem is typically stated as a minimum weight design problem, for all trusses that satisfy a static equilibrium within certain constraints on stresses in the individual structural members. With the same stress constraint value Q_b for both tension and compression, the formulation is in the form of an LP. Cheng and Jiang (1992) and Kirsch (1993a) outline that the stress constraints are written in terms of structural member forces in order to give a consistent formulation. They further state that, for some truss problems, the stress in several structural members will converge to a finite non-zero level as the structural member areas converge to zero, but the structural member forces will converge to zero. This fact should be observed for any truss design problem involving stress constraints (Cheng and Jiang 1992, Kirsch 1993a).

Just as in Bendsøe (2009), let q_b^+, q_b^-, σ_b denote the structural member forces' tension and compression and stress constraint value σ_b for both tension and compression, respectively, so that designs for which all structural members with non-zero member area have stresses at the maximum allowed level Q_b (fully stressed designs) occur. Furthermore,

$$t_b = \frac{L_b}{Q_b}(q_b^+ + q_b^-)\,, \quad \text{with } q = (q^+ - q^-)\,, \quad q^+, q^- \geq 0 \qquad (4.3)$$

applies, so that one can write the problem TTO$_{p;a}$ also as an LP, see problem TTO$_{p;b}$. The equivalence of the problems TTO$_{p;a}$ and TTO$_{p;b}$ is described in Dorn (1964), with a discussion on how the force formulations are convenient for studying static determinacy of solutions. The objective function of the problem TTO$_{p;b}$ is the weight of the truss structure. As reported by Kirsch and Rozvany (1993), a basic solution to the problem TTO$_{p;b}$ encompasses the existence of a minimum mass truss topology with the number of structural members not exceeding the degrees of freedom. If there exists such a basic solution with only non-zero forces (areas), this is a statically determined truss; otherwise, the truss will have a unique force field for the given load and will be kinematically indeterminate (Bendsøe 2009). Problem TTO$_{p;a}$ can be extended to a multiple load LP ($TTO_{p;c}$, see Table 4.1) using a set of $k = \{1, \ldots, s\}$ different loading cases where the self-weight loads are considered. Thus, the sum

$$\sum_{b \in T} t_b g_b \,, \tag{4.4}$$

with g_b denoting the specific nodal gravitational force vector due to self-weight and external loads must be considered, see Equations (4.7b) and (4.8b).

4.2.2   Kinematic Compatibility

The validation and verification method of our algorithm-driven product design process is an FEA, and therefore we are able to distinguish strictly between rigid-body equilibrium of forces calculated via a (quantified) MILP and a verification of the results via a linear elastic and a nonlinear elastic numerical analysis (FEA) of our optimized truss-like structures.

For this reason, in accordance with Bendsøe (2009), it is helpful that the problems $TTO_{p;a}$, $TTO_{p;b}$ and $TTO_{p;c}$ are at first problems in plastic design, as the kinematic compatibility is ignored and their use in elastic design is justified by the possibility of finding statically determinate solutions. In addition, the listed problems could be solved using sparse, primal-dual interior point LP-methods or the simplex algorithm (Bendsøe 2009). We exploit the advantages of both methods. On the one hand, we will focus on efficiently solving large-scale TTO problems with practical relevance for AM with the help of standard solvers; on the other hand, we will verify our results via a linear elastic and nonlinear elastic numerical analysis, which is in any case part of an FEA driven product design process (Kurowski 2006).

Kirsch (1989b, 1993a) and Topping (1992) state it to be well known, that most commonly statically indeterminate solutions result in the case of a multiple load plastic design (problem $TTO_{p;c}$). If kinematic compatibility is required, a further redesign with an FEA becomes necessary in addition to the in any case required FEA validation for elastic design. This further redesign to satisfy kinematic compatibility for elastic design leads to high effort or costs or both, leading to problem $TTO_{p;c}$ being non-practical. Problem $TTO_{p;b}$ is modified and extended by geometry-based design rules for AM, resulting in the MILPs $TTO_{l;p}$ and $TTO_{l;s}$, see Section 5.2 and 5.3. The MILP $TTO_{l;m}$ (see Section 5.4) and QMIP $TTO_{l;q}$ (see Section 5.5) are independent of the TTO models stated in Table 4.1.

Table 4.1 TTO models

Denotation	Model
TTO_c	$$\inf_{t} \min_{q} \frac{1}{2} \sum_{b \in T} \frac{L_b^2}{E_b} \frac{(q_b)^2}{t_b} \qquad (4.5a)$$ $$\text{s.t. } Bq = p \qquad (4.5b)$$ $$\sum_{b \in T} t_b = V \qquad (4.5c)$$ $$t_b > 0 \qquad \forall b \in T \quad (4.5d)$$
$TTO_{p;a}$	$$\min_{q,t} \sum_{b \in T} t_b \qquad (4.6a)$$ $$\text{s.t. } Bq = p \qquad (4.6b)$$ $$-\sigma_b t_b \leq L_b q_b \leq \sigma_b t_b \qquad \forall b \in T \quad (4.6c)$$ $$t_b \geq 0 \qquad \forall b \in T \quad (4.6d)$$
$TTO_{p;b}$	$$\min_{q^+,q^-} \sum_{b \in T} \frac{L_b}{\sigma_b}(q_b^+ + q_b^-) \qquad (4.7a)$$ $$\text{s.t. } B(q^+ - q^-) = p \qquad (4.7b)$$ $$q_b^+ \geq 0, q_b^- \geq 0 \qquad \forall b \in T \quad (4.7c)$$
$TTO_{p;c}$	$$\min_{q^k,t} \sum_{b \in T} t_b \qquad (4.8a)$$ $$\text{s.t. } Bq^k = p^k + \sum_{b \in T} t_b g_b \qquad \forall k = 1, \ldots, s \quad (4.8b)$$ $$-\sigma_b t_b \leq L_b q_b^k \leq \sigma_b t_b \qquad \forall b \in T$$ $$\qquad \forall k = 1, \ldots, s$$ $$t_b \geq 0 \qquad \forall b \in T \quad (4.8d)$$

4.3   Preparations Towards a Truss Design Problem

Our MILP and QMIP models use the so-called static Timoshenko member theory (Timoshenko et al. 1962) for constant cross-sections; see Figure 4.5. We assume a structural member to be a structure, which has one of its dimensions much larger than the other two dimensions. It follows, that the cross-sections of a structural member do not deform in a significant manner under the influence of transverse or axial loads, and therefore the structural member is assumed to be rigid. Besides, allowing no transverse shearing forces and bending moments, the displacement functions depend on the coordinates along the central axis of the structural member. The only allowed loads are external line, and area-loads modeled as concentrated forces acting on connection nodes. We claim a linear elastic isotropic material, with the given deformation restrictions causing no transverse stresses to occur.

4.3.1   Structural Member

We will consider E_b and I_b as the Young's modulus and the area moment of inertia of a structural member $b = \{i, j\}$, respectively. We will denote by κ_b, L_b, A_b, G_b Timoshenko's shear coefficient (Cowper 1966), the length, the cross-sectional area and the shear modulus (Roylance 2008) of a structural member b, respectively. The Timoshenko member theory is equivalent to the Euler-Bernoulli member theory, so that the Inequality (4.9) applies. Using the Timoshenko member theory

$$\frac{E_b I_b}{\kappa_b L_b^2 A_b G_b} \ll 1 \,, \tag{4.9}$$

$$\frac{\partial E_b}{\partial \psi_b} = \frac{\partial G_b}{\partial \psi_b} = 0 \,, \tag{4.10}$$

$$\frac{\partial A_b}{\partial \psi_b} = \frac{\partial I_b}{\partial \psi_b} = 0 \,, \tag{4.11}$$

applies, whereby ψ_b is the central axis of a structural member $b = \{i, j\}$. The basis for Equation (4.10) is formulated based on Hooke's law (Rychlewski 1984), which states a linear dependency between stress and strain. In reality, the linear part is the elastic part of the material property that can be described by the Young's modulus E_b. The shear modulus G_b is defined as the ratio of shear stress to shear strain. The area moment of inertia/second area moment I_b (Gibson 2016) is a geometrical property of an area which reflects how its points are distributed with regard to an arbitrary axis. It is used to determine the deflection and stress caused by a moment applied to a structural member. Equation (4.10) forces a homogeneous material and Equation (4.11) forces a constant cross-section, neglecting manufacturing inaccuracies of AM processes.

Timoshenko's shear coefficient κ_b relates to the Poisson's ratio μ_b (Lakes 1987), which is the negative of the ratio of signed transverse strain to signed axial strain. Concerning these works, we may assume κ_b to be constant. The shear coefficient κ_b is set to

$$\kappa_b = \frac{6(1 + \mu_b)}{7 + 6\mu_b} = \frac{9}{10} \tag{4.12}$$

for a circular cross-section A_b. The physical assumptions of the Timoshenko member theory have to be pre-processed, see Inequality (4.9) and Equations (4.10) to (4.12), as they cannot be added explicitly to the MILP and QMIP models, which can only contain linear constraints unlike a NLP. This is possible since Inequality (4.9) and Equations (4.10) to (4.12) only depend on the dimension and material properties of a structural member. Therefore, dimension and material of a structural member are input parameters of our MILPs, see Sections 5.2 and 5.3.

4.3.2 Force Equilibrium

Let $u \in \mathbb{R}^d$ and $K(A, I) \in \mathbb{R}^{d \times d}$ denote the vector of nodal displacements, where d are the numbers of translatoric degrees of freedom of displacements and the stiffness matrix respectively. The vectors of the cross-sectional area A_b and area moment of inertia I_b of a structural member $b = \{i, j\}$ are written as $A = (A_b \mid b \in E)$ and $I = (I_b \mid b \in E)$. The external load is stated as $f \in \mathbb{R}^d$. The formulation of the equilibrium Equation (4.13) and the decomposition into the force-balance Equation (4.14) along with the following conclusions refer to Kureta and Kanno (2014). The equilibrium equation

$$K(A, I)u = f \tag{4.13}$$

can be decomposed into the force-balance equation and written as

$$\sum_{b \in E} \sum_{l=1}^{3} s_{b,l} \boldsymbol{b}_{b,l} = f \tag{4.14}$$

and the relation between the generalized stresses and the displacement vector is written as

$$s_{b,l} = k_{b,l} \boldsymbol{b}_{b,l}^{\top} \boldsymbol{u}, \quad l = 1, 2, 3; \quad \forall b \in E. \tag{4.15}$$

Here, $\boldsymbol{b}_{b,1}, \boldsymbol{b}_{b,2}, \boldsymbol{b}_{b,3} \in \mathbb{R}$ ($\forall b \in E$) are constant vectors with respect to the local coordinate system of a structural member $b = \{i, j\}$. Constants $k_{b,1}, k_{b,2}, k_{b,3} \in \mathbb{R}$ ($\forall b \in E$) are defined by

$$k_{b,1} = \frac{E_b A_b}{L_b^0}, \tag{4.16}$$

$$k_{b,2} = \frac{1}{L_b} \left(\frac{L_b^2}{12 E_b I_b} + \frac{1}{\kappa_e G_b A_b} \right)^{-1}, \tag{4.17}$$

$$k_{b,3} = \frac{E_b I_b}{L_b^0}, \tag{4.18}$$

where L_b^0 is the undeformed length of a structural member $b = \{i, j\}$. Note that the coefficients $k_{b,1}, k_{b,2}, k_{b,3} \in \mathbb{R}$ are determined by the material selected for a structural member, so that we can preprocess Inequality (4.9) along with Equations (4.10) to (4.12) and (4.16) to (4.18). In line with Kureta and Kanno (2014), it is defined that $k_{b,2} = 0$ if $E_b = G_b = 0$, see Equation (4.17).

By elimination of the displacement vector $s_{b,i}$, Equations (4.14) and (4.15) revert to Equation (4.13). The decomposition into the force-balance Equation (4.14) is a fundamental concept of our MILP and QMIP models, since we want to specify the equilibrium of forces as a hard constraint and the cost per structural member type as a soft constraint, resulting from the pre-processed structural member volume and material specification. A hard constraint is selected as a constraint for the variables that are required to be satisfied. By implication, a soft constraint is a constraint that is allowed to be violated. The force-balance Equation (4.14) can be used in combination with the binary variables x_{ij} (MILPs TTO$_{1;p}$ and TTO$_{1;s}$) or x_e (MILP TTO$_{1;m}$ and QMIP TTO$_{1;q}$) to indicate whether a structural member is present between nodes $i, j \in V$ or at edge $e \in E \subseteq V \times V$; see Constraints (5.8b) to (5.8d) and (5.8g) in the MILP TTO$_{1;p}$, Constraints (5.11b) to (5.11d) and (5.11g) in the MILP TTO$_{1;s}$, Constraints (5.12b) and (5.12e) in the MILP TTO$_{1;m}$, and Constraints (5.14c) and (5.14f) in the QMIP TTO$_{1;q}$. The binary variable $x_{i,j}$ is used to implement further geometry-based construction and design standards (ISO/ASTM 52921, VDI 3405-3-4, VDI 3405-3-3) in the MILP $TTO_{l;s}$, see Section 4.5.

4.3.3 Moment Equilibrium

Let m_k and m_l denote two moments in Newton metre acting centrically at two adjacent nodes, where $k, l \in V$. The lever arms of the moments m_k and m_l are positioned in relation to the global coordinate system $F^G_{W P_{x,y,z}}$ of the reference volume $\mathbb{V}$, as described in Section 4.4 and shown in Figure 4.4. Let $T = \{0, 1, \ldots, n\}$ denote a set of different structural member types t and $F_i j$ the force vector in Newton between the directly adjacent nodes $i \in V$ and $j \in V$. As stated in Inequality (4.19), we assume a linear elastic strains plastic behavior, with the plastic section modulus $Z_b = \infty$ of a structural member $b = \{i, j\}$. The stress $c_{t,i,j}$ of a structural member b of type t in Pascal is stated as

$$\frac{|F_i j|}{A_b} + \overbrace{\frac{max\{|m_k|, |m_l|\}}{Z_b}}^{\approx 0} \le c_{t,i,j} . \tag{4.19}$$

The upper bound of the stress $\bar{c}_{t,i,j}$, affected by the safety factor $\gamma \in]1, +\infty[$, is given by the equation

$$\bar{c}_{t,i,j} = \frac{c_{t,i,j}}{\gamma} . \tag{4.20}$$

4.4 Design Space

4.4.1 Additive Manufacturing Constraints

As can be seen in Figure 4.4, in relation to the front side of the AM system an upwardly directed manufacturing is postulated. The build origin $(0, 0, 0)$ is placed on the origin of the machine coordinate system and thus coincides with the origin of the assembly space $\mathbb{A}$. The build direction Z coincides with the positive y-direction of the global coordinate system. The build surface is located on the x,z-plane and corresponds to the spanned plane of the assembly space $\mathbb{A}$ on the x,z-plane. It is assumed that the build platform of the AM system is sufficiently dimensioned and the feed region and overflow region (see, e.g., Figure 2.13 for the basic principle of

SLS) allow the planned reference volume $\mathbb{V}$. The build platform of the AM system is located in the negative y-direction.

The layer thickness is known as constant, fully dense, and near net shape, so that the accuracy is assumed as sufficient to assume perfect structural members, see Equations (2.29)–(2.31). The pre-processing of the upper bound of the stress $\bar{c}_{t,i,j}$ is therefore realistic, see Cases (5.5). According to ISO/ASTM 52921, the right-hand rule for positive rotations, in relation to the origin of the reference volume $\mathbb{V}$, is applied. All dimensions between the three bounding boxes are known. The assembly space $\mathbb{A}$, also known as master bounding box, is determined by the used AM system and is therefore independent of the truss orientation. The initial build orientation is assumed to be as shown in Figure 4.4. A truss reorientation is permissible so that no restrictions arise at component (truss) nesting (Canellidis et al. 2013). The reference volume $\mathbb{V}$ can change as a result of a truss reorientation, so we assume the reference volume $\mathbb{V}$ to be an arbitrary oriented bounding box. Figure 4.4 illustrates the difference between the reference volume $\mathbb{V}$ and the ground structure $\mathbb{G}$, although in some cases $\mathbb{G} \equiv \mathbb{V}$ may occur.

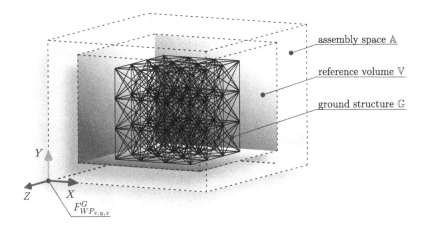

Figure 4.4 Division of the design space

4.4.2 Assembly Space

Since AM is a layered manufacturing process, the build orientation of a truss towards the assembly space has a significant impact on the truss quality (e.g., accuracy and surface finish) and costs (see, e.g.,Allen and Dutta 1994, Frank and Fadel 1995, Pham et al. 1999, and references therein). Also, the support structure depending on the support contact area and the build time are affected. In order to be able to define the direction of our truss-like structure towards the assembly space, we need to make some assumptions. The terminology is derived from ISO/ASTM 52900.

Let $\mathbb{A}$ be the assembly space of an AM system, represented by a polyhedron as an open subset of $\mathbb{R}^3$. Let $\mathbb{V} \subset \mathbb{A}$ be the reference volume (discretized design domain) to be replaced by a ground structure $\mathbb{G}$. The classification of the assembly

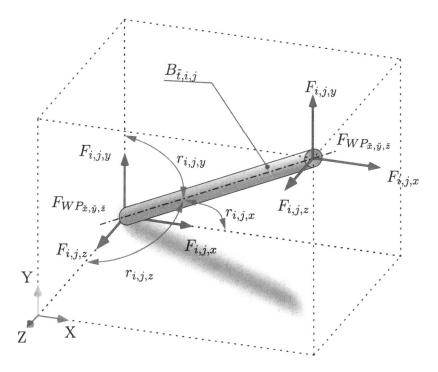

Figure 4.5 Structural member $B_{\tilde{t},i,j}$ and related local coordinate systems $F_{WP_{\tilde{x},\tilde{y},\tilde{z}}}$ and $F_{WP_{\tilde{x},\tilde{y},\tilde{z}}}$

space $\mathbb{A}$ enables us to distinguish between internal and external support structures. Any type of support structure within $\mathbb{V}$ is defined as an internal support structure. A support structure outside of $\mathbb{V}$, but forced to be connected to the load-bearing structure, is defined as an external support structure. Besides, the classification of the assembly space $\mathbb{A}$ is done for minimization of $\mathbb{V}$, in order to save computing effort and to optimize the utilization of the assembly space $\mathbb{A}$. The classification is the formal basis for component (truss) decomposition, in detail orientation decisions for single- and multi-components and packing problems, see Oh et al. (2018).

4.4.3 Ground Structure

Following our aim to develop MILP and QMIP models, which can be applied to large-scale TTO for AM in consideration of geometry-based design rules and structural symmetry, we focus on a ground structure with low complexity. Therefore, only directly adjacent nodes as possible structural member connections are considered. In the MILPs $TTO_{l;p}$ and $TTO_{l;s}$ the angles $r_{i,j,x}, r_{i,j,y}, r_{i,j,z}$ (see Table 5.2) between two possible structural members in the primary planes of a local coordinate system $F_{WP_{x,y,z}}$ are fixed to $45°$; see Figure 4.5. In the MILP $TTO_{l;m}$ and QMIP $TTO_{l;q}$ the set of vertices V and set of edges $E \subseteq V \times V$ (see Tables 5.5 and 5.7) comply with the ground structure in Figure 4.6. Consequently, the ground structure in Figure 4.6 is not the set of all possible connection nodes of a highly connected ground structure.

According to Bendsøe (2009), this approach may obviously lead to designs that are not the best ones for the chosen set of connection nodes, but the approach implicitly allows for restrictions on the possible spectrum of structural member lengths as well as for the study of the optimal subset of structural members of a given truss layout. In consideration of our assumptions, three structural member lengths (see Equation (5.1) and Relation (5.2)) arise, which is the minimum for a three-dimensional truss-like structure with diagonal and bracing structural members. Different ground structures can be set up by adjusting the angles $r_{i,j,x}, r_{i,j,y}, r_{i,j,z}$ (MILPs $TTO_{l;p}$ and $TTO_{l;s}$) or pre-processing a different ground structure (MILP $TTO_{l;m}$ and QMIP $TTO_{l;q}$). Recesses can be created by setting $x_{i,j} = 0$ (MILPs $TTO_{l;p}$ and $TTO_{l;s}$) or $x_e = 0$ (MILP $TTO_{l;m}$ and QMIP $TTO_{l;q}$) for the desired volume.

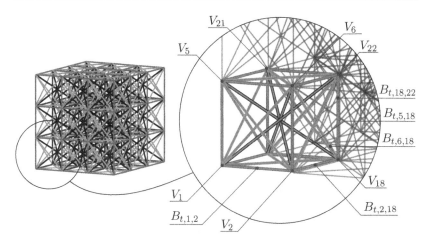

Figure 4.6 An illustration of the ground structure approach using the notation of the MILPs TTO$_{l;p}$ and TTO$_{l;s}$. Note that an arbitrary section of the ground structure is shown in the detail view and not a unit cell structure used to build up the ground structure periodically. Each structural member is considered an individual element

4.5   Support Structure Optimization

According to Jiang et al. (2018), the reason for using support structures (see Figure 2.14) is to ensure manufacturability in the manufacturing process, since a three-dimensional component with overhanging, hole, or edge features will need support structures for successful manufacturing, as printed materials will not be able to stand in position, due to the layer-wise fabrication and the associated stair-stepping effect (Quan et al. 2015). The main support methods for AM processes are honeycomb support, sparse tree support, tree-like support, space-efficient branching support, bridge support, and grain support (Jiang et al. 2018).

In addition to the physical necessity for support structures, depending on the AM process and the complexity of the component geometry, the support structures increase material consumption, manufacturing costs, and post-processing effort. Post-processing can have a negative impact on the component quality or even prevent manufacturability and should be minimized or excluded. Using support structures, a tool-less and post-processing free manufacturing is not possible. An optimization of the support structures would lead to a reduction in the material consumption during the manufacturing process and a significant reduction of the time and effort required during post-processing (Lindecke et al. 2018). Removability has to be considered, especially for an internal support structure. If the internal support structure is irremovable because of no direct access to the support structure, which likely occurs for truss-like structures manufactured with AM, the support structure has to be considered in the component design and validation as force-bearing. Also, a self-manufacturability requirement is necessary, as the support structure itself should not require a support structure (Jiang et al. 2018). In line with Langelaar (2016), existing topology optimization approaches do not consider AM-specific limitations during the design process, resulting in designs that are not self-supporting.

The main categories of methods minimizing the need for support structures for AM can be divided into two subgroups; either the original component is kept intact or the original component is redesigned; see Figure 4.7. We extend the main categories of support structure methods for AM by a global optimization in terms of a support-free component, see Section 5.3.

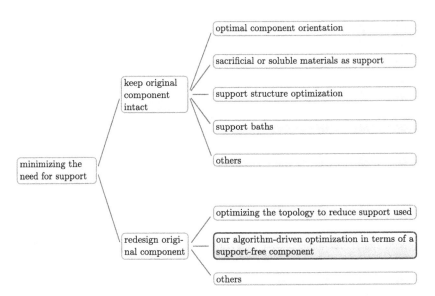

Figure 4.7 Main categories of support structure methods for AM inclusive our algorithm-driven optimization in terms of a support-free component, based on (Jiang et al. 2018)

Research in the field of support structures for AM has primarily focused on reducing the print time, characterization, and optimization of the support materials used and post-processing (Jiang et al. 2018). Vaidya and Anand (2016) presented a new approach for minimizing support structures in conjunction with Dijkstra's shortest path algorithm to generate optimized support structures. They subsequently used a numerical model to prove that the optimized support can withstand the planned loading case. Vanek et al. (2014) proposed a tree-like support structure generation method for Fused Deposition Modeling (FDM), which can reduce manufacturing time and costs. As in the previous method, a numerical analysis is necessary because the method is geometry-based and considers the angle and length of the support structure. The geometry-based optimization and the numerical verification of the stability are strictly separated. Our algorithm-driven optimization depends on component design and manufacturability and it does not require a numerical model to prove that the optimized support can withstand the planned loading case. Any feasible solution is manufacturable and implies no post-processing effort if support is impermissible. We consider the support structure as a subset of the load-bearing truss-like structure and optimize both structures in a single task.

Mixed-Integer Linear Programming for Truss Optimization

5

Chapter 5 is divided into five sections. Section 5.1 presents preliminary work for our MILP and QMIP models. A MILP specifying truss-like structures for powder-based AM processes ($\text{TTO}_{1;p}$), a MILP defining support-free truss-like structures ($\text{TTO}_{1;s}$), and a MILP for optimized and manufacturable cross-sectional areas ($\text{TTO}_{1;m}$) are presented in Section 5.2 to 5.4. Section 5.5 describes the QMIP $\text{TTO}_{1;q}$ for Robust Truss Topology Optimization (RTTO). Some of the ideas presented in this chapter result from intensive discussions with Hartisch (2016–2020). The following formulations for the mathematical optimization models can also be found in the publications (Reintjes and Lorenz, 2019, 2020, 2021, Reintjes et al., 2018, 2019, 2020).

5.1 Preliminaries

Characteristics of AM are tool-less manufacturing, the elimination of geometric restrictions in general, and the paradigm shift from production-ready to function-oriented design (Hague et al., 2003, Langelaar, 2016). The tool-less manufacturing is given a priori by the AM process, whereby we serve the other two characteristics with our MILP and QMIP models. On the one hand, the function-oriented design is obtained by the objective function of the models, while on the other hand, the elimination of geometric restrictions is obtained by variable structural member lengths based on the ground structure, along with the possibility of different structural member cross-sections and diameters. The fundamental phenomenon of AM,

Supplementary Information The online version contains supplementary material available at https://doi.org/10.1007/978-3-658-36211-9_5.

C. Reintjes, *Algorithm-Driven Truss Topology Optimization for Additive Manufacturing*, https://doi.org/10.1007/978-3-658-36211-9_5

„complexity for free", is to be fully exploited by the objective function, to achieve a holistic design approach and overcome conventional manufacturing constraints, see Jared et al. (2017).

Using AM for truss-like structures, no standardized shaped components or connections, and consequently the associated costs, are necessary. For this reason, assembly costs are not taken into account in our models, as a connection node only consists of an accumulation of cured material. If our models are formulated for a classical construction application, the influence of standard fasteners and time differences during the assembly can be considered by additional constraints.

A formulation of the equilibrium of forces as a soft constraint and cost per structural member type as a hard constraint is not practicable for our models. Due to the assumptions of Section 4.3, the potential energy stored in all members, given by the mechanical compliance, would be a soft constraint. We avoid a formulation of the mechanical compliance, since we distinguish strictly between rigid-body equilibrium of forces calculated via a MILP or QMIP and a verification of the results via linear elastic and nonlinear elastic numerical analysis of our optimized truss-like structures.

We set up all models in terms of structural member forces for single load, plastic design. All models are stated as a minimum weight design problem, for all trusses that satisfy static equilibrium within certain constraints on the stresses in the individual members. The same stress constraint value is valid for both tension and compression.

LP versus MILP

Clearly, if all binary variables and the corresponding constraints in $TTO_{l;p}$ (see Section 5.2) or $TTO_{l;s}$ (see Section 5.3) are omitted, the result is one of the LPs TTO_c, $TTO_{p;a}$, $TTO_{p;b}$, $TTO_{p;c}$, of Table 4.1. By omitting the binary variable $x_{i,j}$ in the MILPs $TTO_{l;p}$ and $TTO_{l;s}$, it is no longer possible to model geometry-based design rules depending on the AM process. Furthermore, it is no longer possible to use pre-processed structural member diameters and materials, so that multi-material truss-like structures could be realized. To implement geometry-based design rules, the problem naturally becomes a (quantified) MILP, and we suspect it to be **NP**-hard.We see no other option than to introduce the binary variable $x_{i,j}$ indicating whether a structural member is present between two adjacent nodes $i \in V$ and $j \in V$ and force $F_{i,j}$ to the target function. In the MILP $TTO_{l;m}$ and QMIP $TTO_{l;q}$, we use a binary variable x_e (see Tables 5.4 and 5.6) to indicate the existence of a structural member at edge $e \in E$ with a specified minimum cross-sectional area A_{min} (see Tables 5.5 and 5.7). In addition, we use a continuous variable a_e (see

Tables 5.4 and 5.6) to specify the additional cross-sectional area of a structural member necessary to withstand all specified loading scenarios.

Objective Function and Design For Manufacturability
The objective value does not reflect the manufacturable truss structure's exact volume: Overlapping parts of the structural members (interferences from clashing bodies and free spaces) at the vertices are not merged but counted multiple times; see Figure 5.1 left. As post-processing of interferences and free spaces at the vertices (intersections of the structural members) is inevitable to obtain ready for manufacturing truss-like structures, see Chapter 6 and Figure 5.1 right, the slightly different component volume given by the objective value poses no problem for mathematical programming but for CAD, CAE, FEA, and AM. We argue that rather than implementing constraints to define interferences from clashing bodies and free spaces in our MILPs and QMIPs, which is suspected to be computationally expensive, it might be more useful to use CAD to post-process the intersections and compute the exact component volume. Using our Ansys SpaceClaim add-in `constructOR`, the post-processing can be done, and the exact component volume can be computed easily within the Ansys Workbench (see Chapter 6).

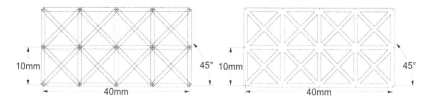

Figure 5.1 Sectional view of a 40-member planar rectangular grid demonstrated in Example 5.1: (left) sketched component volume inclusive interferences from clashing bodies according to the objective value of our MILP and QMIP models; (right) sketched component volume after post-processing of interferences from clashing bodies and free spaces at the vertices using our Ansys SpaceClaim add-in `constructOR`

Undeformed Length of a Structural Member
It is ensured via pre-processing that

$$L_b^0 = N(t(b)) \cdot \tau \tag{5.1}$$

is fulfilled, wherein $N(t(b))$ is the number of layers needed for a structural member b of type t manufactured with the layer thickness τ. The undeformed length of a structural member L_b^0 is given by

$$L_b^0 \in \left\{ \hat{L}_b^0, \frac{\sqrt{2}}{1}\hat{L}_b^0, \sqrt{(\frac{\sqrt{2}}{1}\hat{L}_b^0)^2 + \hat{L}_b^{02}} \right\} =$$

$$\left\{ \hat{L}_b^0, \sqrt{2}\hat{L}_b^0, \sqrt{3}\hat{L}_b^0 \right\} , \tag{5.2}$$

with $\hat{L}_b^0$ being the length of a structural member manufactured purely horizontal or vertical in relation to the coordinates of an AM system. All possible undeformed lengths of a structural member are a multiple of the layer thickness τ, which reduces production inaccuracies.

5.2   The MILP TTO$_{l;p}$ for Powder-Based Additive Manufacturing

Pre-processing the MILPs TTO$_{l;p}$ and TTO$_{l;s}$

According to Equation (4.20), the safety factor γ modifies the upper bound of the capacity $c_{t,i,j}$ of a structural member $\{i, j\}$ of type t. The force components in every direction in space $Q_{i,x}, Q_{i,y}, Q_{i,z} \in \mathbb{R}_+$, resulting from an applied concentrated force Q_i at a node $i \in V$ are decomposed, as described in Subsection 4.3.2. The angles between two possible members in the primary planes of the local coordinate system $F_{WP_{x,y,z}}$ are fixed to $45°$, so that $r_{i,j,x}, r_{i,j,y}, r_{i,j,z} \in \{0, \frac{1}{\sqrt{2}}, 1\}$, the ground structure and the critical upskin and downskin angle $\upsilon_{cr} = \delta_{cr} = 45°$, as described in Section 5.3, applies.

Concerning the preparations towards a truss-like structure design problem as MLIP, see Section 4.3, the costs of a structural member of type t, $cost_t \in \mathbb{R}_+$, are pre-processed and proportional to the volume and material consumed. Moreover, the different capacities $c_{t,i,j} \in \mathbb{R}_+$ of a structural member type t are pre-processed, see Cases (5.5). Let $c_{t,i,j} \in \mathbb{R}_+$ be the pre-processed upper bound of the stress of a structural member type t between the nodes $i \in V$ and $j \in V$ with the associated material constitution

$$M_t = (\kappa_t, L_t, A_t, E_t, G_t, I_t) \tag{5.3}$$

and dimension

$$D_t = (A_t, L_t) . \tag{5.4}$$

We define $\overline{F}_{t,i,j} \in \mathbb{R}$, to be the pre-processed upper bound of the axial force $F_{t,i,j}$ between the nodes $i, j \in V$ depending on the structural member b of type t. The set $U \subseteq E$ is defined as the set of void structural members. The following generic cases–to clarify the pre-processing–are differentiated:

$$
\bar{c}_{t,i,j} = \begin{cases} \overline{F}_{t,i,j}, & \{i, j\} \in E \setminus U, \\ 0, & \{i, j\} \in U. \end{cases} \tag{5.5}
$$

Furthermore, Constraints (5.6) and (5.7) determine the positions $L_{i,x}$, $L_{i,y}$ of a node $i \in V$. The positions $L_{i,x}$, $L_{i,y}$ are measured from the global coordinate system $F^{G}_{W P_{x,y,z}}$ ($L_{1,x} = L_{1,y} = 0$) aligned with the axis x, y, z of the directions in space. We define the length, width, and height of the reference volume $\mathbb{V}$ stated as the number of connection nodes in each direction in space to be $\bar{x}, \bar{y}, \bar{z} \in \mathbb{N}$. The positions $L_{i,x}$, $L_{i,y}$ are used to determine the lever arms of the global statical area moment, see Constraints (5.8e) and (5.8f).

$$
L_{i,x} = (i \mod(\bar{x}\bar{z} - 1)) \mod \bar{x} \qquad \forall i \in V \tag{5.6}
$$

$$
L_{i,y} = \left\lfloor \frac{i-1}{\bar{x}\bar{z}} \right\rfloor \qquad \forall i \in V \tag{5.7}
$$

Instance Parameter of the MILPs TTO$_{l;p}$ and TTO$_{l;s}$

The input parameters of both MILPs are the loading case (Q_i, $i \in V$). The sizing of the reference volume $\mathbb{V}$ is determined by the parameters $\bar{x}, \bar{y}, \bar{z} \in \mathbb{N}$, which are the length, width, and height of the assembly space $\mathbb{V}$, stated as the number of connection nodes in each direction in space. A set of connection nodes $V = \{1, \ldots, \overline{xyz}\}$ results. Taking into account the undeformed length of a structural member $L_b^0 \in \left\{ \hat{L}_b^0, \sqrt{2}\hat{L}_b^0, \sqrt{3}\hat{L}_b^0 \right\}$ (see Equation (5.1) and Relation (5.2)) and the binary variable $B_{t,i,j}$ indicating if a structural member b of type t is present at $\{i, j\} \in E$, the dimensions of the reference Volume $\mathbb{V}$ are given by $B_{t,i,j}$ and $\bar{x}, \bar{y}, \bar{z} \in \mathbb{N}$.

A set of different structural member types $T = \{0, 1, \ldots, n\}$ and a set of nodes V acting as bearings $B \subseteq V$ are determined. The position of a bearing in the reference volume $\mathbb{V}$ corresponds to the indexing of the node i. The maximum permissible load corresponds to the maximum permissible load of the bearing for a static loading case. Analogous to the external forces Q_i the bearing reaction forces R_i are decomposed into the force components $R_{i,x}$, $R_{i,y}$, $R_{i,z}$ with respect to the local coordinate system $F_{W P_{x,y,z}}$ and inserted centrically at a node. The variables used in the models TTO$_{l;p}$ and TTO$_{l;s}$ are given in Table 5.1. The parameters and sets of the models are listed in Tables 5.2 and 5.3, respectively.

Table 5.1 Variables of the MILPs $TTO_{l;p}$ and $TTO_{l;s}$

Symbol	Definition
$F_{i,j} \in \mathbb{R}$	Axial force between nodes $i \in V$ and $j \in V$
$R_{i,x}, R_{i,y}, R_{i,z} \in \mathbb{R}$	Bearing reaction force component in the x, y, or z direction in space concerning the local coordinate system $F_{WP_{x,y,z}}$ inserted centrically at node $i \in V$
$x_{i,j} \in \{0, 1\}$	Binary variable indicating whether a member is present between nodes $i \in V$ and $j \in V$
$y_i \in \{0, 1\}$	Binary variable indicating the minimal and maximal amount of members at node $i \in V$ to comply with the design rules for inclined and support-free cylinders
$B_{t,i,j} \in \{0, 1\}$	Binary variable indicating whether a member of type $t \in T$ is present between nodes $i \in V$ and $j \in V$
$Z_i \in \{0, 1\}$	Binary variable indicating whether at least one support structure constraint is satisfied at node $i \in V$
$\ell_i^o \in \{0, 1\}$	Binary variable identifying all cases that require an additional member to comply with the design rules for inclined and support-free cylinders at node $i \in V$

Table 5.2 Parameters of the MILPs $TTO_{l;p}$ and $TTO_{l;s}$

Symbol	Definition
$c_{t,i,j} \in \mathbb{R}_+$	Capacity of a possible member of type $t \in T$ present between nodes $i \in V$ and $j \in V$
$M \in \mathbb{R}_+$	Big M—maximum capacity of the most robust member type
$cost_t \in \mathbb{R}_+$	Cost of member type $t \in T$
$Q_{i,x}, Q_{i,y}, Q_{i,z} \in \mathbb{R}$	Force component in the x, y, or z direction in space of the external concentrated load Q_i at node $i \in V$
$L_{i,x}, L_{i,y} \in \mathbb{N}$	Level number measured from the reference node $i = 1$ in the direction in space x or y
$\overline{x}, \overline{y}, \overline{z} \in \mathbb{N}$	Length, width, and height of the reference volume $\mathbb{V}$ stated as the number of connection nodes in each direction in space
$r_{i,j,x}, r_{i,j,y}, r_{i,j,z} \in \{0, \frac{1}{\sqrt{2}}, 1\}$	The angles between two possible members in the primary planes x, y, or z of a local coordinate system $F_{WP_{x,y,z}}$, caused by a member structure between adjacent nodes $i \in V$ and $j \in V$

Table 5.3 Sets of the MILPs TTO$_{l;p}$ and TTO$_{l;s}$

Symbol	Definition
$NB(i) \subseteq V$	Adjacent nodes of i
$NB_x(i), NB_y(i), NB_z(i) \subseteq NB(i)$	Adjacent nodes of i which have a force component in the x, y, or z direction in space
$NB'(i) \subseteq NB(i)$	Adjacent nodes of i which can require an additional member to comply with the design rules for inclined and support-free cylinders
$B \subseteq V$	Nodes acting as bearings
$V = \{1, \ldots, \overline{xyz}\}$	Connection nodes of the used ground structure representing the reference volume $\mathbb{V}$
$T = \{0, 1, \ldots, n\}$	Different member types
$O = \{0, 1, \ldots, m\}$	Member combinations between two adjacent nodes that are excluded in a support-free truss-like structure
$A_2 = \{0, 1, \ldots, 8\}$	Possible members at a node within a two-dimensional reference volume $\mathbb{V} \subseteq \mathbb{R}_+^2$
$A_3 = \{0, 1, \ldots, 26\}$	Possible members at a node within a three-dimensional reference volume $\mathbb{V} \subseteq \mathbb{R}_+^3$

The Model TTO$_{l;p}$

$$\min \sum_{i \in V} \sum_{j \in V} \sum_{t \in T} B_{t,i,j} cost_t \tag{5.8a}$$

$$\text{s.t.} \sum_{j \in NB_x(i)} F_{i,j} \cdot r_{i,j,x} + Q_{i,x} + R_{i,x} = 0 \qquad \forall i \in V \tag{5.8b}$$

$$\sum_{j \in NB_y(i)} F_{i,j} \cdot r_{i,j,y} + Q_{i,y} + R_{i,y} = 0 \qquad \forall i \in V \tag{5.8c}$$

$$\sum_{j \in NB_z(i)} F_{i,j} \cdot r_{i,j,z} + Q_{i,z} + R_{i,z} = 0 \qquad \forall i \in V \tag{5.8d}$$

$$\sum_{\substack{i \in V \\ Q_{i,z} \neq 0}} Q_{i,z}(L_{i,y} - L_{j,y})$$

$$+ \sum_{k \in B} R_{k,z}(L_{k,y} - L_{j,y}) = 0 \qquad \forall j \in B \tag{5.8e}$$

$$\sum_{\substack{i \in V \\ Q_{i,z} \neq 0}} Q_{i,z}(L_{i,x} - L_{j,x})$$

$$+ \sum_{k \in B} R_{k,z}(L_{k,x} - L_{j,x}) = 0 \qquad\qquad \forall j \in B \qquad (5.8\text{f})$$

$$F_{i,j} \leq M x_{i,j} \qquad\qquad \forall i, j \in V \qquad (5.8\text{g})$$

$$F_{i,j} = -F_{j,i} \qquad\qquad \forall i, j \in V \qquad (5.8\text{h})$$

$$F_{i,j} \leq \sum_{t \in T} c_{t,i,j} B_{t,i,j} \qquad\qquad \forall i, j \in V \qquad (5.8\text{i})$$

$$B_{t,i,j} = B_{t,j,i} \qquad\qquad \forall i, j \in V, t \in T \quad (5.8\text{j})$$

$$\sum_{t \in T} B_{t,i,j} = x_{i,j} \qquad\qquad \forall i, j \in V \qquad (5.8\text{k})$$

$$R_{i,z} = 0 \qquad\qquad \forall i \in V \setminus B \qquad (5.8\text{l})$$

$$x_{i,j}, B_{t,i,j} \in \{0, 1\} \qquad\qquad \forall i, j \in V, t \in T$$
$$\qquad\qquad\qquad\qquad\qquad\qquad\qquad\qquad (5.8\text{m})$$

The objective function of the MILP $\text{TTO}_{\text{l;p}}$ (5.8a) is to create a statically determined truss-like structure, which is minimal in terms of costs under the influence of external forces. The minimization of costs in the target function represents a minimization of the required volume and material, since these sizes are decisive for the costs. Constraints (5.8b) to (5.8d) determine a force equilibrium at every node $i \in V$ between the decomposed external forces $Q_{i,x}$, $Q_{i,y}$, $Q_{i,z}$ and the active forces $F_{i,j}$ of the structural members mounted at a node i, following the decomposed force-balance Equation (4.14). It is possible to connect each node $i \in V$ to a set of adjacent nodes $NB_x(i)$, $NB_y(i)$, $NB_z(i) \subseteq NB(i)$ which have a force component in the x, y, or z direction in space. The decomposition of the forces $F_{i,j}$ in the directions in space x, y, and z is obtained by applying the appropriate trigonometric values $(r_{i,j,x}, r_{i,j,y}, r_{i,j,z})$ in relation to the local coordinate system $FWP_{x,y,z}$ of the node i.

Constraints (5.8e) and (5.8f) define a global statical area moment which results from the decomposed external forces $Q_{i,x}$, $Q_{i,y}$, $Q_{i,z}$ and the decomposed bearing reaction forces $R_{i,x}$, $R_{i,y}$, $R_{i,z}$, following the assumptions of Equation (4.19) that we assume a linear elastic strains plastic behavior, with the plastic section modulus $Z_b = \infty$ of a structural member $b = \{i, j\}$. Only vertical forces are exerted. Thus, only two moment equilibrium constraints are needed and $R_{i,x} = R_{i,y} = 0$ $\forall i \in V$ applies.

Constraint (5.8g) ensures that only applied structural members ($B_{t,i,j} = x_{i,j} = 1$) can experience active forces. Constraint (5.8h) represents the force equilibrium of the adjacent nodes $i \in V$ and $j \in V$ connected with one structural member indicated by $B_{t,i,j} = 1$. Constraint (5.8i) limits the active force $F_{i,j}$ in the structural member $B_{t,i,j}$ concerning the capacity $c_{t,i,j} \in \mathbb{R}_+$, whereas Constraint (5.8k) guarantees that a specific structural member b of type t is selected if a structural member $B_{t,i,j}$ is used. Combining Constraints (5.8j) and (5.8k) we get $x_{i,j} = x_{j,i}$ $\forall i, j \in V$ which demand a used structural member to withstand tension and compression, which in combination with (5.8h) corresponds to Newton's third law of motion. Finally, Constraint (5.8l) sets the bearing reaction forces of non-bearing nodes to zero. Any node can be defined as a bearing.

5.3   The MILP TTO$_{l;s}$ for Support-Free Truss-Like Structures

Design options to ensure manufacturability for self-supporting cylindrical components, with different geometric characteristics, are presented. Numerical key figures are defined and can be used to engineer support-free components and ensure manufacturability. The geometric characteristics are taken from the standard VDI 3405-3-4 and the terminology is derived from ISO/ASTM 52900.

Optimization-Friendly Additive Manufacturing Constraints

Sectional Shape A varying sectional shape of a cylindrical component is unbounded, as the quality of a metallurgically-bonded connection is independent of the sectional shape. The transition angle β is also unbounded. Due to different mass accumulations, a varying sectional shape can influence the dimensional accuracy.

Minimal Horizontal Manufactured Self-Supporting Cylinder In the case of a minimal horizontal manufactured self-supporting cylinder, dimensional deviations due to the stair-step effect have to be taken into account. For this reason, the layer thickness τ is decisive for designing a horizontal self-supporting cylinder.

Minimal Vertical Manufactured Self-Supporting Cylinder In the case of a minimal vertical manufactured self-supporting cylinder, the contour path including filling has to be ensured so that the track width is decisive. Let D, D_h, D_v be the outer diameter of a self-supporting cylinder independent of the component orientation and dependent on the horizontal/vertical orientation, respectively. For a minimum horizontal manufactured self-supporting cylinder, the outer diameter D_h is set to the value range $\underline{D_h} \gg 2\underline{\tau}$ and $\underline{D_h} \geq 4.0\,mm$. For a minimum vertical manufactured self-supporting cylinder $\underline{D_v} \geq 4.0\,mm$ is defined.

Overhang b_o A self-supporting overhang b_o specifies the overhang of a component to the build platform. A maximum self-supporting overhang $\overline{b}_o$ must not be exceeded, other than that a support structure is necessary. The maximum width of the self-supporting overhang b_o indicates the extent to which no support structure is needed without manufacturing faults occurring. The value for a maximum self-supporting overhang is $\overline{b}_o \leq 1.6\,mm$. In the case of unknown boundaries for the maximum width of the self-supporting overhang b_o, it is recommended to comply with $\overline{b}_o \leq 2.5\,mm$.

Minimum Gap Size The minimum gap size describes the required gap width b_g for immovable components as a function of the surrounding cured material. It is imperative to exclude merging of the adjacent layers. The minimum vertical gap is set to $\underline{b}_{g,v} \geq 0.2\,mm$ and the minimum horizontal gap to $\underline{b}_{g,h} \geq 0.3\,mm$. It is recommended to arrange a gap as vertically as possible in order to avoid support structures, since the stair-step effect is insignificant for vertical component orientation.

Upskin and Downskin Area $\mathbb{U}$ **and** $\mathbb{D}$ An upskin area $\mathbb{U}$ is a (sub-)area whose normal vector in relation to the build direction Z is positive; see Figure 5.2. Similarly, a downskin area $\mathbb{D}$ is a (sub-)area whose normal vector in relation to the build direction Z is negative. The length of a cylindrical structural member is defined as l.

Upskin and Downskin Angle υ **and** δ An upskin angle υ is an angle between the build platform plane and an upskin area $\mathbb{U}$ whose value lies between $0°$ and $90°$. A downskin angle δ is an angle between the build platform plane and a downskin area $\mathbb{D}$ whose value lies between $0°$ (parallel to the build platform) and $90°$ (perpendicular

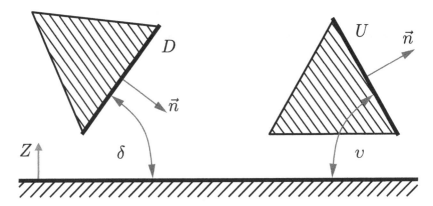

Figure 5.2 Downskin- and upskin angle δ and υ according to standard VDI 3405-3-4

to the build platform); see Figure 5.2. If a perpendicular normal vector exists in relation to the build direction Z ($\upsilon = \delta = 90°$), the upskin area $\mathbb{U}$ and downskin area $\mathbb{D}$ are identical. Both areas coincide with the positive build direction Z and no support structure is needed.

Minimum Angle of Inclined Self-Supporting Cylinders The minimum angle of inclined self-supporting cylinders is limited by the downskin angle δ, not by the upskin angle υ (VDI 3405-3-3). Inclined cylinders can be manufactured without a support structure as long as the critical downskin angle $\delta_{cr} = 45°$ is not undershot. Depending on the downskin angle δ, the manufactured self-supporting cylinder can deform during manufacturing from a certain ratio of overhang/member length $l = L_b^0$ of a cylinder to the outer diameter D if no support structure is provided. Manufacturing faults may occur. The permissible L_b^0/D ratio depends on δ.

Implementation of Optimization-Friendly Additive Manufacturing Constraints Figure 5.3 shows a definition by cases for supporting a cylinder depending on the length $l = L_b^0$, diameter D, downskin angle δ, and the use of support structures. The case distinction is implemented in the MILP TTO$_{l;s}$ by Equations (5.9) and (5.10). Due to the fitted ground structure, $\delta_{cr} = \upsilon_{cr} = 45°$ applies implicitly. It is not necessary to differentiate between the critical downskin angle δ_{cr} and the critical upskin angle υ_{cr} of a support-free cylinder. A component not taking into account the design constraints for support-free cylinders mentioned in this section is classified as not ready for manufacturing. We assume a component to be connected directly to the build platform without an external support structure. We define a maximum

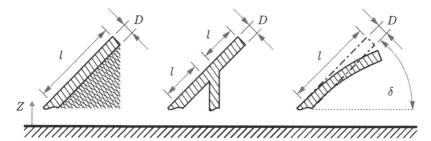

Figure 5.3 (Left) Cylinder with any downskin angle δ and adverse L_b^0/D ratio. Suitable by the support structure, but not optimal in terms of material consumption and post-processing; (Middle) Cylinder suitable for material extrusion AM with minimal manufacturable downskin angle $\delta = \delta_{cr} = 45°$ and admissible L_b^0/D ratio; (Right) Cylinder not suitable for material extrusion AM with $\delta \geq \delta_{cr}$ and too large L_b^0/D ratio. According to standard VDI 3405-3-3

length $\bar{l} = \bar{L}_b^0$ of a cylindrical structural member depending on the AM process so that the curl effect (VDI 3405-3-3) is excluded. A modification of internal holes to avoid support structures (Thomas, 2009) is unnecessary, since the predefined ground structure applies and thus all internal holes or free spaces or both are known.

Even though a varying sectional shape of a cylindrical component is unbounded, we assume $\underline{D}_h = \underline{D}_v \geq 4.0\,mm$ to minimize dimensional accuracy and fulfill the design options for a minimum horizontal and vertical manufactured self-supporting cylinder. A change in the cross-sectional area is fixed to a transition angle of $\beta = 90°$, so that only vertical transitions between cylindrical components are possible, independent of the component orientation. The value for a maximum self-supporting overhang is set to $\bar{b}_o = 1.6\,mm$. We fix the minimum horizontal and vertical gap size to $\underline{b}_{g,h}$, $\underline{b}_{g,v} \geq 0.3\,mm$. The build direction Z is assumed to be positive with increasing node indexing, so that a distinction between upskin and downskin areas $\mathbb{U}$ resp. $\mathbb{D}$, upskin and downskin angles υ resp. δ, is possible. By fixing the critical upskin and downskin angle to $\upsilon_{cr} = \delta_{cr} = 45°$ via the ground structure and introducing a permissible length l to diameter D ratio of a cylinder, the following two cases as implementation of the standards VDI 3405-3-4 and VDI 3405-3-3

$$\upsilon_{cr} = \delta_{cr} = 45° : L_b^0/D \leq 5\,, \tag{5.9}$$

$$\upsilon = \delta = 90° : L_b^0/D \leq 10\,, \tag{5.10}$$

follow. Note that, these design rules are only enforced locally for each edge. In particular, if multiple structural members with identical central axes form one long structural member, then post-processing is necessary. Constraint (5.9) identifies a cylinder manufactured angled at $45°$ in relation to the local coordinate system $F_{WP_{x,y,z}}$ of a node. Analogous to that, Constraint (5.10) identifies a cylinder manufactured angled at $90°$ in relation to the local coordinate system $F_{WP_{x,y,z}}$ of a node. Constraints (5.9) and (5.10) restrict any combination of single or multiple cylinders with identical structural member axes set in the ground structure. With the above discussion, we now state the model $TTO_{l;s}$ for $k = \{2, 3\}$:

The Model $TTO_{l;s}$

$$\min \sum_{i\in V}\sum_{j\in V}\sum_{t\in T} B_{t,i,j}\, cost_t \tag{5.11a}$$

$$\text{s.t.} \sum_{j\in NB_x(i)} F_{i,j}^x + Q_{i,x} + R_{i,x} = 0 \qquad \forall i \in V \tag{5.11b}$$

$$\sum_{j \in NB_y(i)} F_{i,j}^y + Q_{i,y} + R_{i,y} = 0 \qquad\qquad \forall i \in V \qquad (5.11c)$$

$$\sum_{j \in NB_z(i)} F_{i,j}^z + Q_{i,z} + R_{i,z} = 0 \qquad\qquad \forall i \in V \qquad (5.11d)$$

$$\sum_{\substack{i \in V \\ Q_{i,z} \neq 0}} Q_{i,z}(L_{i,y} - L_{j,y})$$
$$+ \sum_{k \in B} R_{k,z}(L_{k,y} - L_{j,y}) = 0 \qquad \forall j \in B \qquad (5.11e)$$

$$\sum_{\substack{i \in V \\ Q_{i,z} \neq 0}} Q_{i,z}(L_{i,x} - L_{j,x})$$
$$+ \sum_{k \in B} R_{k,z}(L_{k,x} - L_{j,x}) = 0 \qquad \forall j \in B \qquad (5.11f)$$

$$F_{i,j} \leq M x_{i,j} \qquad\qquad \forall i,j \in V \qquad (5.11g)$$

$$F_{i,j} = -F_{j,i} \qquad\qquad \forall i,j \in V \qquad (5.11h)$$

$$F_{i,j} \leq \sum_{t \in T} c_{t,i,j} B_{t,i,j} \qquad\qquad \forall i,j \in V \qquad (5.11i)$$

$$B_{t,i,j} = B_{t,j,i} \qquad\qquad \forall i,j \in V, t \in T \quad (5.11j)$$

$$\sum_{t \in T} B_{t,i,j} = x_{i,j} \qquad\qquad \forall i,j \in V \qquad (5.11k)$$

$$2\ell_i^o \leq \sum_{j \in NB^\backslash(i)} x_{i,j} \leq \ell_i^o + 1 \qquad \forall i,j \in V, o \in O \quad (5.11l)$$

$$\sum_{j \in NB(i)} x_{i,j} \geq 3Z_i \qquad\qquad \forall i \in V \qquad (5.11m)$$

$$\min(A_k)y_i \leq \sum_{j \in NB(i)} x_{i,j} \leq \max(A_k)y_i \qquad \forall i,j \in V \qquad (5.11n)$$

$$\sum_{o \in O} \ell_i^o \leq \frac{1}{2}\max(A_k)Z_i \qquad\qquad \forall i \in V \qquad (5.11o)$$

$$R_{i,z} = 0 \qquad\qquad \forall i \in V \setminus B \qquad (5.11p)$$

$$x_{i,j}, y_i, \ell_i^o, Z_i, B_{t,i,j} \in \{0,1\} \qquad\qquad \forall i,j \in V, t \in T, o \in O$$
$$(5.11q)$$

The objective function of the MILP TTO$_{l;s}$ (5.11a) is to create a statically determined, load-bearing, and support-free truss-like structure, which is minimal in terms of cost. The minimization of cost in the target function represents a minimization of the required volume and material following the MILP $TTO_{l;p}$. The Constraints (5.11l) to (5.11o) present a geometry-based modeling, affecting the force equilibrium Constraints (5.8b) to (5.8d) and the moment equilibrium Constraints (5.8e) and (5.8f). The force and moment equilibrium constraints are bounded by Constraints (5.8g) to (5.8i).

Constraints (5.9) and (5.10) apply due to the implementation of Constraints (5.11l) to (5.11q). Let O denote the set of structural member combinations between two adjacent nodes that are excluded in the MILP $TTO_{l;s}$ in opposition to $TTO_{l;p}$ to realize a support-free truss-like structure. Constraints (5.11l) and (5.11o) identify the combination possibilities of structural members between adjacent nodes $NB^\backslash(i)$ that would not comply with the design rules of VDI 3405-3-4 and VDI 3405-3-3. Constraint (5.11l) checks all adjacent nodes $NB^\backslash(i)$, which can require an additional structural member to comply with the design rules for support-free truss-like structures. All cases that require an additional structural member to comply with the design rules are identified by the binary variable ℓ_i^o in Constraint (5.11l). Associated in Constraint (5.11o), the binary variable Z_i indicates whether at least one additional structural member is needed. Constraint (5.11m) forces the model to add at least one structural member between two adjacent nodes identified as critical in Constraint (5.11l) and (5.11o). Constraint (5.11n) forces the model to set the binary variables $x_{i,j}$ and y_i, which corresponds to the construction of an additional structural member to comply with the design rules for support-free truss-like structures.

5.4   The MILP $TTO_{l;m}$ for Manufacturable Cross-Sectional Areas

In order to formally represent the ground structure (see Figure 4.6) an undirected graph $G = (V, E)$ is used with vertices (frictionless joints) and connecting edges (straight and prismatic structural members). Additionally, a set of bearings $B \subseteq V$ must be specified. Note that the vertices are fixed in space, as angles between two possible structural members and distances between joints matter in our modeling approach. We additionally require that the resulting truss-like structure is symmetric with respect to two symmetry planes; see Figure 5.4. We use the function R : $E \to E$, mapping edge e to its representative $R(e)$ in order to enforce that the structural members at edges e and $R(e)$ share the same cross-sectional area. Due to manufacturing restrictions a structural member must have a minimum cross-sectional area. Therefore, we use a binary variable x_e to indicate the existence of a structural member at edge $e \in E$ with a specified minimum cross-sectional area and a continuous variable a_e to specify its additional cross-sectional area. The continuous

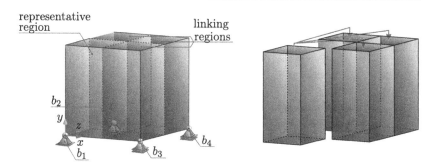

Figure 5.4 Bearing positions and symmetry: Mapping edge e to its representative $R(e)$ in order to enforce that the structural members at edges e and $R(e)$ share the same cross-sectional area

variable n_e represents the normal force in a structural member at edge e and r_b specifies the bearing reaction force of bearing b. The variables, sets, and parameters used in the MILP TTO$_{l;m}$ are given in Table 5.4 and Table 5.5, respectively. We use bold letters when referring to a vector or matrix. The external forces are given by **F**.

The Model TTO$_{l;m}$

$$\min \sum_{e \in E} L_e \left(A_{\min} \cdot x_{R(e)} + a_{R(e)} \right) \tag{5.12a}$$

$$\text{s.t. } S|n_e| \le \sigma_y \left(A_{\min} \cdot x_{R(e)} + a_{R(e)} \right) \qquad \forall\, e \in E \tag{5.12b}$$

$$\sum_{e \in I(b)} n_e^d + F_b^d + r_b^d = 0 \qquad \forall\, b \in B,\, d \in \{x, y, z\} \tag{5.12c}$$

Table 5.4 Variables of the MILP TTO$_{l;m}$

Symbol	Definition
$\mathbf{x} \in \{0, 1\}^E$	x_e: indicator, whether a member is present at edge e
$\mathbf{a} \in \mathbb{R}_+^E$	a_e: additional cross-sectional area of a member e
$\mathbf{r} \in \mathbb{R}^{B \times 3}$	r_b^d: bearing reaction force at bearing b in spatial direction $d \in \{x, y, z\}$
$\mathbf{n} \in \mathbb{R}^E$	n_e: normal force in a member present at edge e

Table 5.5 Sets and parameters of the MILP TTO$_{l;m}$

Symbol	Definition
V	set of vertices
$E \subseteq V \times V$	set of edges
$I : V \rightarrow 2^E$	$I(v) = \{e \in E \mid v \in e\}$: set of edges incident to vertex v
$B \subseteq V$	set of bearings
$L_e \in \mathbb{R}_+$	length of edge e
$A_{\min} \in \mathbb{R}_+$	minimum cross-sectional area of a member
$A_{\max} \in \mathbb{R}_+$	maximum cross-sectional area of a member
$\sigma_y \in \mathbb{R}_+$	yield strength of the cured material
$S \geq 1$	factor of safety
$\mathbf{F} \in \mathbb{R}^{V \times 3}$	F_v^d: external force at vertex v in spatial direction $d \in \{x, y, z\}$
$\mathbf{V}(v, v') \in \mathbb{R}^3$	vector from $v \in V$ to $v' \in V$ (corresponding to lever arm)
$R : E \rightarrow E$	$R(e)$: edge representing edge e due to symmetry

$$\sum_{e \in I(v)} n_e^d + F_v^d = 0 \qquad\qquad\qquad \forall v \in V \backslash B, d \in \{x, y, z\}$$

$$\text{(5.12d)}$$

$$a_{R(e)} \leq (A_{\max} - A_{\min}) x_{R(e)} \qquad\qquad \forall e \in E \qquad \text{(5.12e)}$$

$$\sum_{v \in V} \mathbf{V}(b, v) \times \mathbf{F}_v + \sum_{b' \in B} \mathbf{V}(b, b') \times \mathbf{r}_{b'} = 0 \qquad \forall b \in B \qquad \text{(5.12f)}$$

$$\sum_{v \in V} \mathbf{F}_v + \sum_{b \in B} \mathbf{r}_b = 0 \qquad\qquad\qquad\qquad \text{(5.12g)}$$

$$\mathbf{x} \in \{0, 1\}^E, \; \mathbf{a} \in \mathbb{R}_+^E, \; \mathbf{r} \in \mathbb{R}^{B \times 3}, \; \mathbf{n} \in \mathbb{R}^E \qquad\qquad \text{(5.12h)}$$

The objective function of the MILP TTO$_{l;m}$ (5.12a) aims at minimizing the costs (volume) of the resulting stable and (symmetric) spatial truss-like structure considering the external static loading case. Constraint (5.12b) ensures that the local longitudinal stress in a structural member must not exceed the structural member's yield strength taking into account a factor of safety. Constraints (5.12c) and (5.12d) ensure the static equilibrium at each vertex of the structure. The decomposition n_e^d of the normal force n_e into each direction in space $d \in \{x, y, z\}$ is obtained by multiplying n_e with $\sin(\theta)$, where θ is the angle between the structural member and the corresponding d-axis. As the ground structure is fixed in space those coefficients can be pre-processed using spatial and angular relationships. Variables indicating an additional cross-sectional area are bound to be zero by Constraint (5.12e) if no structural member is present.

Constraints (5.12f) and (5.12g) define the equilibrium of moments by resolution of the external forces and ensure, in combination with Constraints (5.12c) and (5.12d), that the resulting structure is always a static system of purely axially loaded structural members. In particular, the cross product $\mathbf{V}(b, v) \times \mathbf{F}_v$ is the moment caused by the external force $\mathbf{F}_v$ on bearing b with lever arm $\mathbf{V}(b, v)$. Analogously, $\mathbf{V}(b, b') \times \mathbf{r}_{b'}$ is the moment about bearing b caused by the bearing reaction force at b'.

5.5   The QMIP TTO$_{l;q}$ for Robust Truss Topology Optimization

RTTO deals with the structural design optimization problem to find a truss that is stable and minimal in volume, subjected to loading uncertainty (Ben-Tal and Nemirovski, 1997). It has been widely investigated in (Gally et al., 2015, Kanno and Guo, 2010, Yonekura and Kanno, 2010).

In contrast to optimizing the worst-case loading scenario over a given design-dependent uncertainty set of occasional loads (Gally et al., 2015), we utilize QMIP in order to optimize a truss-like structure for a set of loading scenarios without having to specify a worst-case loading scenario. Our approach considers sizing and topology optimization simultaneously and utilizes the ground structure (see Figure 4.6) which is given by a set of vertices (frictionless joints) and a set of edges (straight and prismatic structural members) of a truss-like structure. The resulting truss-like structure is ensured to be adequately dimensioned for every loading scenario with minimal volume. Therefore, the challenging task of identifying, analyzing, and quantifying the worst-case scenario (Kanno and Guo, 2010), which thus far depended highly on having a practical engineering experience, can be bypassed.

In structural mechanics, symmetry is often exploited to effectively optimize and analyze structural systems (Marsden and Ratiu, 2013). From the viewpoint of mathematical optimization, enforcing symmetry is often computationally beneficial as it leads to a reduction in the number of free variables (see Observations 3.1 and 3.2 and the subsequent discussion). Generally, however, demanding symmetry results in the generation of sub-optimal solutions, as it is possible for optimal solutions to be asymmetric although the design domain, the external loads, and the boundary conditions are symmetric in nature (Stolpe, 2010). We optionally consider two vertical planes of symmetry, as shown in Figure 5.4, hence, specifying the structure of the representative region suffices.

Robust Truss Toplogy Optimization

Our robust model aims at *finding a spatial truss-like structure with minimal volume such that, for any anticipated loading scenario, the structure remains in a trivial state of equilibrium.* This can be restated in the following quantification context:

$$\exists \text{ structure } \quad \forall \text{ loading scenarios } \quad \exists \text{ static equilibrium.} \qquad (5.13)$$

This indicates a question: Does a truss-like structure *exist* such that, *for all* anticipated loading scenarios, a static equilibrium also *exists*. Within the ground structure (see Figure 4.6), given by the undirected graph $G = (V, E)$, the edges must be selected where straight and prismatic structural members should be placed, and for each structural member, a corresponding cross-sectional area must be determined. The locations of the vertices, i.e., nodal points, are fixed in space, to allow the pre-processing of the spatial and angular relationships between edges and vertices. Additionally, a set of bearings $B \subseteq V$ must be specified; see Figure 5.4. However, we also want to be able to demand symmetric structures, and therefore, we introduce the function $R : E \to E$, for which $R(e)$ maps to the edge representing e. In contrast to the established design-variable linking technique (Rao, 2019), only variables that characterize representative structural members need to be deployed to enforce that structural members located at the edges—e and $R(e)$—are equally dimensioned. We use $R(e) = e$ if no symmetry is demanded.

As a minimum cross-sectional area is essential owing to the manufacturing restrictions, we have used the combination of a binary variable x_e and a continuous variable a_e to indicate the existence of a structural member at the edge $e \in E$ (with specified minimum area) and its potential additional cross-sectional area. The minimum area $A_{\min}$ can either be a fixed value or optionally adhere to design rules VDI 3405-3-3, VDI 3405-3-4. In the latter case, the minimum area of each structural member is computed separately, depending on its length and spatial orientation, cf. Section 5.3. Note that, these design rules are only enforced locally for each edge. In particular, if multiple structural members with identical central axes form one long structural member, then post-processing is necessary.

Both $\mathbf{x}$ and $\mathbf{a}$ are the first-stage existential variables as they represent the selected structure. Binary universal variables $\mathbf{y}$ are used to specify the occurring loading scenarios: A single binary universal variable y_i indicates whether *loading case i* is active or not, while an assigned variable vector $\mathbf{y}$ highlights the selected *loading scenario*. For each anticipated loading scenario, the structure given by $\mathbf{x}$ and $\mathbf{a}$ must have the following properties: Each nodal point as well as the entire structure must be in a static equilibrium position, and the longitudinal stress within each structural member—induced by the normal force per cross-sectional area—must not exceed its yield strength. The existential variables n_e and r_b represent the normal force in a structural member at e and the bearing reaction force at bearing b, respectively. The variables used in the QMIP TTO$_{l:q}$ have been showcased in Table 5.6, and the parameters are listed in Table 5.7. We use bold letters to refer to a vector or matrix.

Table 5.6 Variables of the QMIP TTO$_{l:q}$

Symbol	Stage	Definition
$\mathbf{x} \in \{0, 1\}^E$	1($\exists$)	x_e: indicates whether a member is present at edge e
$\mathbf{a} \in \mathbf{R}_+^E$	1($\exists$)	a_e: additional cross-sectional area of a member e
$\mathbf{y} \in \{0, 1\}^C$	2($\forall$)	y_i: indicates whether loading case s is active
$\mathbf{r} \in \mathbf{R}^{B \times 3}$	3($\exists$)	r_b^d: bearing reaction force at bearing b in spatial direction $d \in \{x, y, z\}$
$\mathbf{n} \in \mathbf{R}^E$	3 ($\exists$)	n_e: normal force in a member present at the edge e

Table 5.7 Sets, parameters, and functions of the QMIP TTO$_{l:q}$

Symbol	Definition
V	set of vertices
$E \subseteq V \times V$	set of edges
$I : V \to 2^E$	$I(v) = \{e \in E \mid v \in e\}$: set of edges incident to vertex v
$B \subseteq V$	set of bearings
$L_e \in \mathbb{R}_+$	length of edge e
$A_{\min} \in \mathbb{R}_+$	minimum cross-sectional area of a member
$A_{\max} \in \mathbb{R}_+$	maximum cross-sectional area of a member
$\sigma_y \in \mathbb{R}_+$	yield strength of the cured material
$S \geq 1$	factor of safety
$C \in \mathbb{N}_+$	number of considered loading cases
$\mathbf{F}_i \in \mathbb{R}^{V \times 3}$	$F_{i,v}^d$: external force at vertex v in direction $d \in \{x, y, z\}$ in loading case i
$\mathbf{V}(v, v') \in \mathbb{R}^3$	vector from $v \in V$ to $v' \in V$ (corresponding to lever arm)
$R : E \to E$	$R(e)$: edge representing edge e due to symmetry

The Model TTO$_{l;q}$

$$\min \sum_{e \in E} L_e \left(A_{\min} \cdot x_{R(e)} + a_{R(e)} \right) \tag{5.14a}$$

$$\text{s.t. } \exists \mathbf{x} \in \{0, 1\}^E \ \mathbf{a} \in \mathbb{R}_+^E \ \forall \mathbf{y} \in \{0, 1\}^C \ \exists \mathbf{r} \in \mathbb{R}^{B \times 3} \ \mathbf{n} \in \mathbb{R}^E: \tag{5.14b}$$

$$S|n_e| \leq \sigma_y \left(A_{\min} x_{R(e)} + a_{R(e)} \right) \qquad \forall e \in E \tag{5.14c}$$

$$\sum_{e \in I(b)} n_e^d + \sum_{i=1}^{C} y_i F_{s,b}^d + r_b^d = 0 \qquad \forall b \in B, \ d \in \{x, y, z\} \tag{5.14d}$$

$$\sum_{e \in I(v)} n_e^d + \sum_{i=1}^{C} y_i F_{s,v}^d = 0 \qquad \forall v \in V \setminus B, \ d \in \{x, y, z\} \tag{5.14e}$$

$$a_{R(e)} \leq (A_{\max} - A_{\min}) x_{R(e)} \qquad \forall e \in E \tag{5.14f}$$

$$\sum_{v \in V} \sum_{i=1}^{C} \mathbf{V}(b, v) \times y_i \mathbf{F}_{s,v} + \sum_{b' \in B} \mathbf{V}(b, b') \times \mathbf{r}_{b'} = \mathbf{0} \qquad \forall b \in B \tag{5.14g}$$

$$\sum_{v \in V} \sum_{i=1}^{C} y_i \mathbf{F}_{s,v} + \sum_{b \in B} \mathbf{r}_b = \mathbf{0} \tag{5.14h}$$

The objective function of the QMIP TTO$_{l;q}$ (5.14a) aims at minimizing the costs (volume) of the truss-like structure. The Quantification Sequence (5.14b) defines the variables' respective domain and order, as outlined in Expression (5.13). If $R(e) \neq e$, $\mathbf{x}$ and $\mathbf{a}$ variables are deployed only for the edges in the image of E under function R, i.e., only for the representative edges. In order to avoid extensive notation, this detail is neglected in the quantification sequence.

Constraint (5.14c) ensures that the local longitudinal stress must not exceed the structural member's yield strength considering the factor of safety. In particular, modifying the cross-sectional area given by $A_{\min} + a_{R(e)}$ alters the stress in a structural member. Constraint (5.14c) is linearized by writing the constraint once with the left-hand side $+n_e$ and once with $-n_e$. Constraints (5.14d) and (5.14e) ensure static equilibrium at each vertex. The decomposition n_e^d of the normal force n_e into different spatial directions $d \in \{x, y, z\}$ is obtained by multiplying n_e with $\sin(\theta)$, where θ is the angle between the structural member and the corresponding d-axis. As the ground structure is fixed in space, those coefficients can be pre-processed. With Constraint (5.14f), a_e will be zero in the case no structural member is present at the edge e. Constraints (5.14g) and (5.14h) define the equilibrium of moments by the resolution of the external forces and ensure, in combination with Constraints (5.14d) and (5.14e), that the resulting truss-like structure is always a static system of purely axially loaded structural members. In particular, the cross product $\mathbf{V}(b, v) \times \mathbf{F}_{s,v}$ characterizes the moment induced by external force $\mathbf{F}_{s,v}$ on bearing b with a lever arm $\mathbf{V}(b, v)$. Analogously, $\mathbf{V}(b, b') \times \mathbf{r}_{b'}$ is the moment associated with bearing b caused by the bearing reaction force at b'.

The resulting truss-like structure is ensured to be adequately dimensioned for each of the 2^C loading scenarios resulting from the combination of C loading cases. However, if the resulting truss-like structure only needs to be adequately dimensioned for each individual loading case, the model can be altered by enforcing $\sum_{i=1}^C y_i = 1$ on the universal variables. In this case, only C loading scenarios, which correspond to the loading cases, are of interest. As such, a constraint cannot simply be added to the constraint system, see Hartisch et al. (2016)—this case is implemented using a single-integer universal variable that specifies the loading case, which is then transformed into existential indicator variables. In order to illustrate our approach, we use a 40-member planar rectangular grid displayed in Figure 5.5.

Example 5.1   (40-member planar rectangular grid).
Consider a 40-member planar rectangular grid (40 mm × 20 mm) with 10 mm basic vertex distance, a fixed bearing at the bottom-left corner, floating bearing at the bottom-right corner, and four anticipated color-coded loading cases, see Figure 5.5. We have assumed four loading cases and are interested in structures that can withstand each of the $2^C = 2^4$ loading scenarios resulting from combining individual loading cases. We examine the optimization results for several values

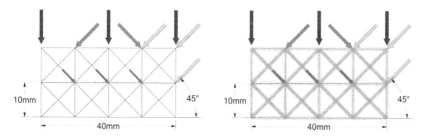

Figure 5.5 Ground structure of a 40-member planar rectangular grid (40 mm × 20 mm) with 10 mm basic vertex distance and all four color-coded loading cases: (left) wire structural members, (right) solid structural members

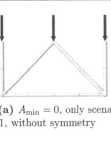

(a) $A_{min} = 0$, only scenario 1, without symmetry

(b) $A_{min} = 0$, only scenario 2, without symmetry

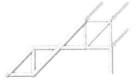

(c) $A_{min} = 0$, only scenario 3, without symmetry

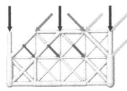

(d) $A_{min} = 0$, only scenario 4, without symmetry

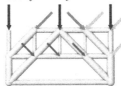

(e) $A_{min} = 0$, holds for each single scenario, without symmetry

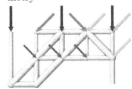

(f) $A_{min} = 0$, holds for all 2^4 scenarios, without symmetry

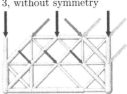

(g) $A_{min} = 0$, holds for all 2^4 scenarios, without symmetry

(h) $A_{min} \equiv A_{VDI}$, holds for all 2^4 scenarios, without symmetry

(i) $A_{min} \equiv A_{VDI}$, holds for each single scenario, without symmetry

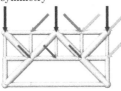

(j) $A_{min} \equiv A_{VDI}$, holds for each single scenario, with symmetry

Figure 5.6 Optimal solutions for the 40-member planar rectangular grid and different loading cases. Figures (h)—(j) are dimensioned as per VDI 3405-3-3 and VDI 3405-3-4 (see Chapter 5.3). The maximum cross-sectional area is always $A_{max} = 10$ mm

of the minimum cross-sectional area $A_{\min}$. The maximum cross-sectional area is always $A_{\max} = 10$ mm. For the sake of simplicity, we do not define all loading scenarios and material characteristics and continue to show our main results. The individual loading cases without symmetry are given in Figures 5.6a–5.6d. Additionally, optimal structures for each individual case have been displayed to ensure better comprehensibility. In Figure 5.6e, the best found robust solution is displayed, which is stable for each single scenario. Figure 5.6f contains the best-found robust solution which is stable in each of the $2^C = 2^4$ loading scenarios. In Figure 5.6g, the optimal solution for $A_{\min} = 0$ mm is displayed, which is stable for any combination of the loading cases. Figure 5.6h contains the optimal solution if each single structural member must be dimensioned as per VDI 3405-3-3 and VDI 3405-3-4. In the case that the loading cases can only occur individually, the optimal solution would be the one shown in Figure 5.6i. Figure 5.6j highlights the optimal solution if we additional demand symmetry around the vertical mid-axis.

Note that, in neither case, one has to explicitly deal with any kind of worst-case scenarios, but that a solution is in a static equilibrium in *every* scenario can be assured. Most importantly, generally, one cannot assume the worst-case scenario to be the one where all loading cases are active since as a compensation of forces might occur.

Figure 5.5 is a first step towards enhancing the understanding of quantified programming and TTO. The solutions of loading cases 5.6a–5.6d are negligible during optimization, which is due to the fact that all loading scenarios are optimized in one step, see, e.g., the robust solution 5.6h. In future instances we often examine the optimization results for several values of $A_{\min}$, but refer to the minimum diameter $D_{\min}$ for the purpose of presentation. In addition, we already refer to the electronic supplementary material on pages 3–4, where more solutions of loading cases 5.6a–5.6d are presented.

CAD-Based Mathematical Optimization 6

With our 3D-CAD tool, we make global optimization of (quantified) MIPs, i.e., the opportunity to find the global solution for a TTO problem with the performance of the state-of-the-art mathematical solvers, available to the CAD and CAE communities. In Section 6.1, we describe the motivation behind and relevance of our 3D-CAD tool as part of our algorithm-driven product design process. In Section 6.2, details on how the 3D-CAD tool is implemented to deal with mathematical optimization instance and solution data are presented, and this is followed by a performance study in Section 6.3. Finally, in Section 6.4, a GUI is presented.

6.1 Motivation

Current design methods and 3D-CAD tools are not tailored to the shape (geometric complexity) of additively manufactured truss-like structures and have not yet been optimized to achieve the great potential offered by AM (see, e.g., Rosen 2013, and the references therein). The innovation in the AM technology has not yet been followed by an adaptation in design and 3D-CAD software and tools (Azman et al. 2014). Furthermore, the truss-like structures' geometric complexity makes the design process in 3D-CAD computationally inefficient (Rosen 2007).

Most commercially available 3D-CAD software use a Boundary REPresentation (BREP) system and are adjusted for shapes manufacturable with conventional instead of AM processes (Azman et al. 2014). The use of exact BREP models enables the repair and preparation of 3D-CAD models with all the manufacturing

Supplementary Information The online version contains supplementary material available at https://doi.org/10.1007/978-3-658-36211-9_6.

C. Reintjes, *Algorithm-Driven Truss Topology Optimization for Additive Manufacturing*, https://doi.org/10.1007/978-3-658-36211-9_6

data included, i.e., ready-for-machine-interpretation 3D-CAD models that can be modified in all the design stages necessary for designing a component. However, this is computationally intensive. As can be seen in Figure 4.3, early triangulation, i.e., converting the mathematical optimization solution data directly into STL data, would reduce the need for computing capacity and decrease the (algorithm-driven) product development cycle time, but the generation of 3D-CAD, CAE, and, partly, CAM data from the mathematical optimization data would be excluded. In addition—contrary to the claims in our algorithm-driven product design process as stated in Subsection 4.1.2—numerical shape optimization and verification of the mathematical optimization data via linear and nonlinear elastic numerical analysis would not be possible. From this, we conclude that an early triangulation to save computing effort is impractical for our algorithm-driven product design process. The 3D-CAD model should remain based on the original geometry description, e.g., a neutral data exchange format like STEP (ISO 10303-242:2020-04), till the data is converted into CAE or CAM or both kinds of data.

For all these reasons, the design process of truss-like structures in 3D-CAD software is highly computationally expensive and limits the number of components (structural members) that can be automatically built in a 3D-CAD model (Rosen 2007, 2013). In addition, most truss-like structures need post-processing of intersections and interferences, geometry cleanup, and simplification to generate (simplified) data that is suitable for an FEA. Hence, transforming large-scale mathematical optimization data, e.g., data exported as a CPLEX instance and solution file, into a 3D-CAD model ready-for-machine-interpretation and suitable for FEA is not a straightforward task at the time of writing. It is necessary to develop 3D-CAD tools (add-ins) for standard 3D-CAD software that are tailored to complex-shaped and topology-optimized additively manufactured truss-like structures.

To overcome the described shortcoming and realize our algorithm-driven product design process, we developed a 3D-CAD tool. It is capable of obtaining well-performing and ready-for-machine-interpretation CAD data from raw mathematical optimization data. It has been implemented twice: in the 3D-CAD software Autodesk Inventor Professional 2020[1] and in Ansys SpaceClaim 2020 R2[2]. The main version is the one implemented in Ansys SpaceClaim, and it is called constructOR and it includes a GUI (see Section 6.4). Both versions are a bi-directional data integration between CPLEX and the respective 3D-CAD software. As can be seen in Figure 4.3,

---

[1] Further information regarding Autodesk Inventor Professional can be found on https://www.autodesk.com/products/inventor/ (accessed February 28, 2021).

[2] Further information regarding Ansys SpaceClaim can be found on https://www.ansys.com/products/3d-design/ansys-spaceclaim/ (accessed February 28, 2021).

the 3D-CAD tool takes the solution file of CPLEX (SOL file) and the instance data of our MILPs and QMIP as input to generate a ready-for-machine-interpretation 3D-CAD model of the optimized truss-like structure in a neutral data exchange format like STEP (ISO 10303-242:2020-04). CAD-based design encodings of the instance data and the truss-like structure are again directed to CPLEX. A finite element software directly adopts the 3D-CAD data and computes (on request) the linear and nonlinear elastic numerical analysis of the optimized truss-like structure. The highly compressed 3D models (e.g., STL, additive manufacturing file format) and CAM data (e.g., machine setting, toolpath generation) are derived based on the generated 3D-CAD design data.

Our 3D-CAD tool takes full advantage of the geometrical freedom of AM by generating each structural member of a truss-like structure individually and from scratch. Therefore, it is possible to generate any type of truss-like structure (see Definitions 2.1–2.3). The 3D-CAD tool can also be used as a stand-alone software independent of the algorithm-driven product design process, i.e., as an Ansys SpaceClaim add-in[3], to generate any type of truss-like structure.

6.2   Implementation Details

We concentrate on a construction methodology (algorithmic procedure) to model any truss-like structure. The established construction methodology to model a unit cell structure, e.g., a all face-centered cubic or body-centered cubic unit cell structure (see, e.g., Bai et al. 2018), that is multiplied in the spatial directions in space, to build up a truss-like structure periodically, is insufficient for this purpose. On the one hand, the construction methodology is not able to model any truss-like structure, i.e., a compound or complex spatial truss-like structure (see Definitions 2.2 and 2.3); on the other hand, a serious criticism is that the computational performance of the construction methodology depends strongly on the geometric complexity of the unit cell structure (Azman et al. 2014).

Our approach is to break down the possible truss-like structure, i.e., the ground structure, into its individual parts and model each structural member from scratch. Figure 6.1 shows an exemplary and notably simplified modeling of a structural member with a circular and constant cross-section using the mathematical optimization instance and solution data of our MILPs $TTO_{1;p}$ and $TTO_{1;s}$ (see Sections 5.2 and 5.3). First, a 2D sketch plane is defined on a connection node $i \in V$ (see Table 5.3),

---

[3] Further information about customizing Ansys SpaceClaim can be found on http://help.spaceclaim.com/2015.0.0/en/ (accessed February 28, 2021).

i.e., a fixed-in-space 3D point[4], as a dimensional reference, and a circular profile (swept arc) is created as a 2D sketch (see Figure 6.1 left). Second, the circular profile in the 2D sketch plane is extruded to the thickness of a structural member to form a volume body with a closed profile (see Figure 6.1 middle and right). The pre-processed, undeformed length of a structural member depending on the layer thickness L_b^0 (see Constraints (5.1) and (5.2)) and its pre-processed dimension D_t (see Constraint (5.4)) are used. The pre-processed material constitution M_t of a structural member of type t (see Constraint (5.3)) is considered.

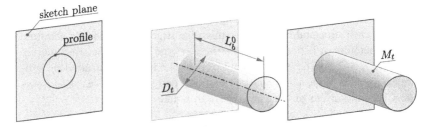

Figure 6.1 Highly simplified construction of a structural member with a circular and constant cross-section

Any truss-like structure can be modeled using this construction methodology, and the computational performance is dependent on the number of components but independent of the geometric complexity of the unit cell structure—in our construction methodology, not in the existing ones—due to the individual construction of each component. The decomposition of the unit cell structure into its individual components is indispensable for us since we cannot make any statement about the type of truss-like structure before mathematical optimization, except to analyze the ground structure, i.e., the upper bound of the possible structural members.

To realize a straightforward implementation of new types of structural members implying different structural member profiles, e.g., hollow structural members, the computation of the reference volume $\mathbb{V}$ and the modeling of a structural member are separated. In Lines 1–4 in Algorithm 1 and Lines 1–5 in Algorithm 2, the implementation of the reference volume $\mathbb{V}$ for Autodesk Inventor Professional 2020 and Ansys SpaceClaim 2020 R2, respectively, is presented.

---

[4] All degrees of freedom of the points in space are removed.

In Lines 5–8 in Algorithm 1 and Lines 1–9 in Algorithm 3, the implementation of the modeling of a structural member as volume body for Autodesk Inventor Professional 2020 is presented. In Lines 6–9 and 10–13 in Algorithm 2, Lines 1–13 in Algorithm 4, and Lines 1–12 in Algorithm 5, the algorithmic procedures of the modeling of a structural member for Ansys SpaceClaim 2020 R2 using the beam class of the Ansys SpaceClaim Application Programming Interface (API) or volume bodies are sketched. We use simple pseudocode notation following the notation given by the Ansys SpaceClaim API[5]. For the sake of simplicity and abbreviation, we just show the algorithmic procedures for the MILPs TTO$_{1;p}$ and TTO$_{1;s}$ for Autodesk Inventor Professional 2020 and Ansys SpaceClaim 2020 R2 and the algorithmic procedure of the post-processing of structural members as volume bodies (see Algorithm 6). The algorithmic procedures for the MILP TTO$_{1;m}$ and the QMIP TTO$_{1;q}$ for Autodesk Inventor Professional 2020 and Ansys SpaceClaim 2020 R2 and the post-processing for beam objects are developed and implemented analogously.

Data Processing of our MILPs TTO$_{1;p}$ and TTO$_{1;s}$
The reference volume $\mathbb{V}$ (see Figure 4.4) is the initial volume considered by our 3D-CAD tool. The reference volume $\mathbb{V}$ is implemented in Lines 1–4 in Algorithm 1 and Lines 1–5 in Algorithm 2, using the parameters $\bar{x}$, $\bar{y}$, $\bar{z} \in \mathbb{N}$ (see Table 5.2) read from the MILP instance file. In Line 7 of Algorithm 1, the binary variable $x_{i,j}$ (see Table 5.1) read from the MILP solution file specifies whether a structural member has to be modeled between the positions in space of the connection nodes i, $j \in V$. In Line 9 of Algorithm 4, Line 1 of Algorithm 5, and Lines 1, 6, and 7 of Algorithm 6, the binary variable $B_{t,i,j}$ (see Table 5.1) read from the MILP solution file indicates whether a structural member of type $t \in T$ with pre-processed material constitution M_t and dimension D_t has to be modeled between the positions in space of the connection nodes i, $j \in V$ and also whether one of the connection nodes i, $j \in V$ has to be post-processed. Note that M_t and D_t are functions of several variables and parameters that are necessary for the generation of the 3D-CAD design data.

---

[5] Further information regarding the Ansys SpaceClaim API be found on http://help. spaceclaim.com/2015.0.0/en/ (accessed February 28, 2021).

The variable $F_{i,j}$ (see Table 5.1) read from the MILP solution file provides the active (axial) force of a structural member between the positions in space of the connection nodes i, $j \in V$, and this allows a comparison between the static structural dimensioning found by our MILPs in combination with the solver CPLEX and linear and nonlinear elastic numerical FEA. The utilization of an individual structural member is defined by the comparison of the capacity $c_{t,i,j}$ (see Table 5.2) of a structural member of type t (independent of i and j) and the active force $F_{i,j}$ of a structural member. The undeformed length L_b^0 and Timoshenko's shear coefficient κ_b of a structural member b of type t apply pre-processed to the instance parameter of the MILP. The cross-sectional area A_t (see Constraint (5.4)), shear modulus G_b, Young's modulus E_b, and the area moment of inertia I_b apply to the material assigned in the 3D-CAD model and given pre-processed in the MILP instance file. Regarding the pre-processing of the above-mentioned parameters, please refer to the discussion starting in the context of preparations for a TTO problem. By considering the data processing mentioned above, numerical analysis and numerical shape optimization can be carried out to validate the mathematical optimization solutions (see Figures 7.9 and 7.10 for an example).

Main Function: Modeling the Reference Volume
One major task when converting the mathematical optimization instance and solution data of our MILPs $\text{TTO}_{l;p}$ and $\text{TTO}_{l;s}$ into a 3D-CAD model according to the 3D-CAD standards is to generate the missing geometric relationships that are indispensable in 3D-CAD for defining the behavior of a drawing object if it gets modified at a later stage in design. To fulfill this task, in the first step, all fixed-in-space work points $WP_{x,y,z}$ (datum points $DP_{x,y,z}$) are created out of raw mathematical optimization data using Lines 1–4 in Algorithm 1 (Autodesk Inventor Professional) or Lines 1–5 in Algorithm 2 (Ansys SpaceClaim). In the second step, the structural members are modeled in relation to the coordinates x, y, z of the fixed-in-space work points $WP_{x,y,z}$ (datum points $DP_{x,y,z}$) and in relation to the origin $F_{WP_{x,y,z}}^G$ of the reference volume $\mathbb{V}$ (see Figure 4.4). Using this algorithmic procedure, structural members can be modeled directly by indexing the fixed-in-space work points or datum points, and we can bypass the need of modeling angular relationships in our MILPs, except those angular relationships necessary for modeling the force and moment equilibrium constraints.

Autodesk Inventor Professional The algorithmic procedure to generate a truss-like structure out of mathematical optimization data using the software Autodesk Inventor Professional and the iLogic functionality (see, e.g., ASCENT 2019, and the references therein) is presented in Algorithm 1. The algorithm is called as the main function, which results in a call of Algorithm 3 in Line 8. First, we iterate over the set of connection nodes $V = \{1, \ldots, \bar{x}\,\bar{y}\bar{z}\}$ (see Tables 5.2 and 5.3), representing the length, width, and height of the reference volume $\mathbb{V}$ as the number of connection nodes in each direction in space (see Lines 1–3 in Algorithm 1). Second, all structural members are modeled after calling Algorithm 3 in Line 8 of Algorithm 1. The set of connection nodes V is implemented as fixed-in-space work points $WP_{x,y,z}$ having the same indexing as the (ordered) index set of connection nodes V. Each work point $WP_{x,y,z}$ is saved in an Autodesk Inventor Professional part file (*.ipt) and contains three coordinates x, y, z in relation to the origin of the reference volume $\mathbb{V}$. The name (identifier) of each work point in the structure tree (ASCENT 2019) in Autodesk Inventor Professional is given by the set V. The distance between two adjacent fixed-in-space work points $WP_{x,y,z}$ arranged on a single plane, i.e., the basic connection node distance, is read from the MILP instance file. All other distances and angular relationships, e.g., to model swept arcs and distances for pull operations (ASCENT 2019), are determined depending on the fixed-in-space work points $WP_{x,y,z}$ using Autodesk Inventor Professional.

The inputs to Algorithm 1 are the MILP instance and solution files and the pre-processed material constitution M_t and dimension D_t of a structural member $B_{t,i,j}$ of type t. The MILP instance file specifies the positions of the fixed-in-space work points $WP_{x,y,z}$ in the reference volume $\mathbb{V}$ and all implicit structural member lengths L_b^0 (see Relation 5.2 and Figure 6.1 middle). The MILP solution file specifies which fixed-in-space work points $WP_{x,y,z}$ are to be connected with the structural member $B_{t,i,j}$ using the binary variable $x_{i,j}$, indicating whether a structural member is present between two adjacent connection nodes implemented as fixed-in-space work points $WP_{x,y,z}$ in Autodesk Inventor Professional. The variable $S_{i,k,d}$ is a function of a connection node $i \in V$ and is used for post-processing interferences from clashing bodies (structural members) and free spaces at a connection node through the generation of a sphere. Regarding the post-processing, please refer to Algorithm 3.

Algorithm 1 returns any truss-like structure as an Autodesk Inventor Professional assembly file (*.iam), with structural members (solid bodies) as Autodesk Inventor Professional part files (*.ipt). All part files are linked to the assembly file.

Ansys SpaceClaim The algorithmic procedure to generate the truss-like structure out of mathematical optimization data using Ansys SpaceClaim is presented in

Algorithm 2. The algorithm is called as the main function, which results in a call of
Algorithm 4 or Algorithm 5 in Lines 7 or 11 of Algorithm 2. Algorithm 6 is called
in Lines 9 or 13 of Algorithm 2 if post-processing of intersections and interferences
at a specific connection node of the truss-like structure is necessary. We iterate
over the set of connection nodes $V = \{1, \ldots, \bar{x}\bar{y}\bar{z}\}$ (see Lines 1–3 in Algorithm
2) representing the length, width, and height of the reference volume $\mathbb{V}$ stated as
the number of connection nodes in each direction in space. Each point in space
$P_{x,y,z}$ (data object) including its location is used to generate a new (visualized)
fixed-in-space datum point $DP_{x,y,z}$ (see Lines 4 and 5 in Algorithm 2). The call
of Algorithm 6 in lines 9 or 13 in Algorithm 2 is used to generate a fixed-in-space
sphere on all datum points $V_{DP} = \{1, \ldots, \bar{x}\bar{y}\bar{z}\}$ that are shared by at least two
structural members that require post-processing.

Algorithm 1: ReferenceVolume, sketched

    Algorithm ReferenceVolume *(instance_ file, solution_ file, M_t, D_t)*

       input : <u>MILP</u>: *instance_ file, solution_ file*, pre-processed material
             constitution M_t and dimension D_t
             <u>Autodesk Inventor</u>:

       output: a truss-like structure as assembly file *(*.iam)*

       `/* create the reference volume V existing of` $V = \{\bar{x}\bar{y}\bar{z}\}$
          `fixed-in-space work points`   `*/`

1   **foreach** $x \in \{1, \ldots, \bar{x}\}$ *in instance_ file* **do**

2   **foreach** $y \in \{1, \ldots, \bar{y}\}$ *in instance_ file* **do**

3   **foreach** $z \in \{1, \ldots, \bar{z}\}$ *in instance_ file* **do**

4               $WP_{x,y,z} :=$ WorkingPoint.Create(x, y, z);

            end

         end

     end

       `/* create all structural members as volume body; necessary`
          `structural members are read from the` <u>MILP</u> `solution file`   `*/`

5   **foreach** $i \in V$ *in instance_ file* **do**

6   **foreach** $j \in NB_i$ **do**

7   **if** $x_{i,j} == 1$ **and** $i < j$; `//` $x_{i,j}$ `read from the` <u>MILP</u> `solution`
               `file`
            then

8   ModelMember$(i, j, B_{t,i,j}, M_t, D_t, S_{i,k,d})$;

            end

         end

     end

    return *Autodesk Inventor Professional assembly file (*.iam)*

There are essentially two strategies for modeling a structural member. First, all structural members are modeled as beam objects using the beam class of the Ansys SpaceClaim API, the library of standard profiles, and the materials library of Ansys SpaceClaim[6] (see Lines 6–9 in Algorithm 2 and Algorithm 4). Second, all structural members are modeled using our algorithmic procedure, which generates all structural members from scratch as volume bodies (see Lines 10–13 in Algorithm 2 and Algorithm 5) by considering only the mathematical optimization data.

The inputs to Algorithm 2 are the MILP instance and solution files, the profile type read from the library of standard beam profiles of Ansys SpaceClaim, the pre-processed material constitution M_t of a structural member $B_{t,i,j}$ of type t read from the MILP instance file, and the binary information if the truss-like structure should be modeled using the beam class of the Ansys SpaceClaim API or our implementation using volume bodies. The MILP instance file specifies the set of locations in space $V_P = \{1, \ldots, \bar{x}\bar{y}\bar{z}\}$ and all possible implicit structural member lengths L_b^0. The MILP solution file, which includes the variable $B_{t,i,j}$, specifies which points $P_{x,y,z}$, i.e., which two adjacent points $P_{\bar{x},\bar{y},\bar{z}}$, $P_{\hat{x},\hat{y},\hat{z}}$, are to be connected with a structural member implemented as design body $DB_{P_{\bar{x},\bar{y},\bar{z}\leftrightarrow\hat{x},\hat{y},\hat{z}}}$ (see Algorithm 5 Line 12 and SpaceClaim Corporation (2019)) or beam object $B_{P_{\bar{x},\bar{y},\bar{z}\leftrightarrow\hat{x},\hat{y},\hat{z}}}$ (see Algorithm 4 Line 13 and SpaceClaim Corporation (2019)). The efficiency of Algorithms 4 and 5 are crucial to the performance of Algorithm 2.

The algorithm returns any truss-like structure as an Ansys SpaceClaim design file (*.scdoc). The design file is instantiated as a root object and top-level design component. Every structural member is a child object of the Ansys SpaceClaim design file.

---

[6] Further information regarding the beam class, the library of standard profiles, and the materials library can be found on http://help.spaceclaim.com/2015.0.0/en/ (accessed February 28, 2021).

Algorithm 2: GenerateTrussLikeStructure, sketched

Algorithm GenerateTrussLikeStructure(*instance_file, solution_file, profile_type, M_t, BeamClass, VolumeBody, identifier*)

> **input** : MILP: *instance_file, solution_file*, pre-processed material constitution M_t
> construcTOR: *profile_type* read from the SpaceClaim library, binary decision to use the *BeamClass* or *VolumeBody*, *identifier*
>
> **output** : a truss-like structure as single top-level design file (**.scdoc*)
>
> /* create the reference volume V existing of $V_P = \{1, \ldots, \bar{x}\bar{y}\bar{z}\}$
>   locations and $V_{DP} = \{1, \ldots, \bar{x}\bar{y}\bar{z}\}$ fixed-in-space datum points   */
>
> 1 **foreach** $x \in \{1, \ldots, \bar{x}\}$ *in instance_file* **do**
> 2   **foreach** $y \in \{1, \ldots, \bar{y}\}$ *in instance_file* **do**
> 3   **foreach** $z \in \{1, \ldots, \bar{z}\}$ *in instance_file* **do**
> 4 $P_{x,y,z} \leftarrow$ Point.Create(x, y, z); // location of the datum point
> 5 $DP_{x,y,z} \leftarrow$ DatumPoint.Create(Document.MainPart, identifier, $P_{x,y,z}$); // new datum point with point as location
>   **end**
>   **end**
> **end**
>
> /* create all structural members given in the solution file; use
>   the beam class or manually create structural members   */
>
> 6 **if** *BeamClass* = *true* **then**
> 7   ModelMemberBC(instance_file, solution_file, profile_type, M_t, identifier); // model structural member using the beam class and a standard beam profile
> 8   **foreach** $v \in V_{DP}$ *in instance_file* **do**
> 9   PostProcessing(instance_file, solution_file, profile_type, identifier);
>   **end**
> **end**
> 10 **if** *VolumeBody* = *true* **then**
> 11   ModelMemberVB(solution_file, M_t, identifier); // model structural member using a volume body
> 12   **foreach** $v \in V_{DP}$ *in instance_file* **do**
> 13   PostProcessing(instance_file, solution_file, profile_type, identifier);
>   **end**
> **end**
> **return** *Ansys SpaceClaim design file (*.scdoc)*

Modeling a Structural Member

Autodesk Inventor Professional Algorithm 3 models a structural member as post-processed volume body using two adjacent work points $WP_{\check{x},\check{y},\check{z}}$, $WP_{\hat{x},\hat{y},\hat{z}}$, the pre-processed structural member profile A_t read from the MILP solution file, and the binary variable $B_{t,i,j}$ read from the MILP solution file to indicate whether a structural member with pre-processed material constitution M_t read from the MILP instance file has to be modeled between the positions in space of the adjacent work points $WP_{\check{x},\check{y},\check{z}}$, $WP_{\hat{x},\hat{y},\hat{z}}$. The variable $S_{i,k,d}$ is used for post-processing. In Lines 1–2 in Algorithm 3, two local coordinate systems (frames) $F_{WP_{\check{x},\check{y},\check{z}}}$, $F_{WP_{\hat{x},\hat{y},\hat{z}}}$ are created centered on two adjacent work points $WP_{\check{x},\check{y},\check{z}}$, $WP_{\hat{x},\hat{y},\hat{z}}$. Both local coordinate systems $F_{WP_{\check{x},\check{y},\check{z}}}$, $F_{WP_{\hat{x},\hat{y},\hat{z}}}$ are modeled in relation to the origin $F^G_{WP_{x,y,z}}$ of the reference volume $\mathbb{V}$. Subsequently, in Line 3, a line segment $L_{WP_{x,y,z}}$ depending on the position in space of the two adjacent work points $WP_{\check{x},\check{y},\check{z}}$, $WP_{\hat{x},\hat{y},\hat{z}}$ is created. The line segment $L_{WP_{x,y,z}}$ is defined as the central axis of a structural member. In Line 4, a plane $P_{WP_{x,y,z}}$ is created as a function of the line segment $L_{WP_{x,y,z}}$ and a work point $WP_{\check{x},\check{y},\check{z}}$. The plane $P_{WP_{x,y,z}}$ is arranged orthogonal to the line segment $L_{WP_{x,y,z}}$. The plane $P_{WP_{x,y,z}}$, as a function of the line segment $L_{WP_{x,y,z}}$ and the work points $WP_{\check{x},\check{y},\check{z}}$, $WP_{\hat{x},\hat{y},\hat{z}}$, is used to implement a construction methodology based only on the MILP instance and solution data: The line segment $L_{WP_{x,y,z}}$ is dependent on the indexing implemented in the MILP, and the plane $P_{WP_{x,y,z}}$ is dependent on the line segment $L_{WP_{x,y,z}}$, except for an angular parameter which defines that the plane $P_{WP_{x,y,z}}$ is orthogonal to the line segment $L_{WP_{x,y,z}}$. In Line 5, a round cross-sectional area $C_{WP_{\check{x},\check{y},\check{z}}}$, as structural member profile centered on one of the two adjacent work points $WP_{\check{x},\check{y},\check{z}}$, $WP_{\hat{x},\hat{y},\hat{z}}$, is created. In Line 6, the structural member profile is extruded along the line segment $L_{WP_{x,y,z}}$ to the thickness of the structural member $B_{WP_{\check{x},\check{y},\check{z}\leftrightarrow\hat{x},\hat{y},\hat{z}}}$ to form a volume body with a closed profile. Next, in Line 7, the material constitution M_t is assigned to each structural member. Finally, in Lines 8–9, we post-process the connection nodes, as displayed in Figures 6.2 and 6.3. Algorithm 3 returns a post-processed structural member as Autodesk Inventor Professional part file (*.*ipt*).

Algorithm 3: ModelMember, sketched

Algorithm ModelMember $(WP_{\hat{x},\check{y},\check{z}},\ WP_{\hat{x},\hat{y},\hat{z}},\ B_{t,i,j},\ M_t,\ A_t,\ S_{i,k,d})$:

    input : <u>MILP</u>: *instance_file, solution_file*, pre-processed material constitution M_t

                <u>Autodesk Inventor:</u>

    output: a structural member as single part file (*.ipt)

        /* create a structural member with circular cross-sectional area;
           merge the topology; add a sphere surface at the intersection
           point; A_t, M_t are read from the MILP instance file   */

1 $F_{WP_{\hat{x},\check{y},\check{z}}} \leftarrow$ Frame.Create$(WP_{\hat{x},\check{y},\check{z}})$;

2 $F_{WP_{\hat{x},\hat{y},\hat{z}}} \leftarrow$ Frame.Create$(WP_{\hat{x},\hat{y},\hat{z}})$;

3 $L_{WP_{x,y,z}} \leftarrow$ Line.CreateThroughPoints$(WP_{\hat{x},\check{y},\check{z}},\ WP_{\hat{x},\hat{y},\hat{z}})$;

4 $P_{WP_{x,y,z}} \leftarrow$ Plane.Create$(WP_{\hat{x},\check{y},\check{z}},\ L_{WP_{x,y,z}})$;

5 $C_{WP_{\hat{x},\hat{y},\hat{z}}} \leftarrow$ Profile.Create$(F_{WP_{\hat{x},\check{y},\check{z}}},\ P_{WP_{x,y,z}},\ L_{WP_{x,y,z}},\ A_t,\ B_{t,i,j})$; // A_t
        read from the MILP instance file

6 $B_{WP_{\hat{x},\check{y},\check{z}\leftrightarrow\hat{x},\hat{y},\hat{z}}} \leftarrow$ Body.ExtrudeProfile$(C_{WP_{\hat{x},\hat{y},\hat{z}}},\ L_{WP_{x,y,z}})$;

7 $B_{WP_{\hat{x},\check{y},\check{z}\leftrightarrow\hat{x},\hat{y},\hat{z}}} \leftarrow$ Body.MaterialProperty$(B_{WP_{x,y,z}},\ M_t)$; // M_t read from
        the MILP instance file

8 $\overline{B}_{WP_{\hat{x},\check{y},\check{z}\leftrightarrow\hat{x},\hat{y},\hat{z}}} \leftarrow$ Body.MergeTopology$(B_{WP_{x,y,z}})$;

9 $\overline{\overline{B}}_{WP_{\hat{x},\check{y},\check{z}\leftrightarrow\hat{x},\hat{y},\hat{z}}} \leftarrow$ Body.Surface.IntersectCurve$(B_{WP_{x,y,z}},\ S_{i,k,d})$;

    return $\overline{\overline{B}}_{WP_{\hat{x},\check{y},\check{z}\leftrightarrow\hat{x},\hat{y},\hat{z}}}$

Ansys SpaceClaim: Beam Class Algorithm 4 models a structural member as a beam object using the beam class of the Ansys SpaceClaim API, the library of standard beam profiles, the pre-processed material constitution M_t read from the MILP instance file, and the structural member profile A_t read from the MILP instance file. The binary variable $B_{t,i,j}$ read from the MILP solution file indicates whether a structural member has to be modeled. In Lines 1–8 in Algorithm 4, all closed beam profiles read from the MILP instance file are generated using the SpaceClaim API without creating a beam. Using this algorithmic procedure, we separated the data processing of the generation of the beam profiles as *.scdoc* file and the generation and visualization of the beams. In Lines 9–13, all beams read from the MILP solution file are generated. In Line 4, we create the vector $V_{P_{\hat{x},\check{y},\check{z}\leftrightarrow\hat{x},\hat{y},\hat{z}}}$ between two adjacent points $P_{\hat{x},\check{y},\check{z}}$, $P_{\hat{x},\hat{y},\hat{z}}$ to get the direction of this vector and create an "arranged-in-space" frame $F_{P_{\hat{x},\check{y},\check{z}}}$ in Line 5. This way, we can generate the missing geometric relationships between two adjacent connection nodes in the MILP instance and solution data. Algorithm 4 returns a structural member as beam object as an element of a truss-like structure saved as a single top-level design file (*.scdoc*).

Algorithm 4: ModelMemberBC, sketched

Algorithm ModelMemberBC (*instance_file, solution_file, profile_type, M_t, identifier*)

    input : <u>MILP</u>: *instance_file, solution_file*, pre-processed material constitution M_t

          <u>constructOR</u>: *profile_type* from SpaceClaim library, *identifier*

    output : a structural member modeled as beam object using the beam class of the Ansys SpaceClaim API, the library of standard profiles, and the materials library of Ansys SpaceClaim; the structural member is an element of a truss-like structure as single top-level design file (*.scdoc)

    /* Create all beam profiles read from the MILP instance file

       using the SpaceClaim API without creating a beam   */

1   **foreach** A_t *in instance_file* **do**

2        $P_{\check{x},\check{y},\check{z}} \leftarrow$ Point.Create($\check{x}, \check{y}, \check{z}$);

3        $P_{\hat{x},\hat{y},\hat{z}} \leftarrow$ Point.Create($\hat{x}, \hat{y}, \hat{z}$);

4        $V_{P_{\check{x},\check{y},\check{z}\leftrightarrow\hat{x},\hat{y},\hat{z}}} \leftarrow P_{\check{x},\check{y},\check{z}} - P_{\hat{x},\hat{y},\hat{z}}$;

5        $F_{P_{\check{x},\check{y},\check{z}}} \leftarrow$ Frame.Create($P_{\check{x},\check{y},\check{z}}$, $V_{P_{\check{x},\check{y},\check{z}}}$.Direction);

6        $P_{P_{x,y,z}} \leftarrow$ Plane.Create($F_{P_{\check{x},\check{y},\check{z}}}$);

7        $PR_{P_{\check{x},\check{y},\check{z}}} \leftarrow$ Profile.Create($P_{P_{x,y,z}}$, A_t, profile_type); // merge A_t

          read from the MILP instance file with the library of

          standard profiles of Ansys SpaceClaim

8        $BP_{P_{\check{x},\check{y},\check{z}}} \leftarrow$ BeamProfile.Create(Window.ActiveWindow.Document, identifier, $PR_{P_{\check{x},\check{y},\check{z}}}$);

    end

    /* Create all structural members as beam objects; necessary

       structural members are read from the MILP solution file;

       pre-processed material constitution from the MILP instance

       file   */

9   **foreach** $B_{t,i,j} = 1$ *in solution_file* **do**

10   $CS_{P_{\check{x},\check{y},\check{z}\leftrightarrow\hat{x},\hat{y},\hat{z}}} \leftarrow$ CurveSegment.Create(new $LineSegment(P_{\check{x},\check{y},\check{z}}$, $P_{\hat{x},\hat{y},\hat{z}})$);

11   $DC_{P_{\check{x},\check{y},\check{z}\leftrightarrow\hat{x},\hat{y},\hat{z}}} \leftarrow$ DesignCurve.Create(Document.MainPart, $CS_{P_{\check{x},\check{y},\check{z}\leftrightarrow\hat{x},\hat{y},\hat{z}}}$);

12   $BP_{P_{\check{x},\check{y},\check{z}\leftrightarrow\hat{x},\hat{y},\hat{z}}} \leftarrow$ Beam.Create($BP_{P_{\check{x},\check{y},\check{z}}}$, $DC_{P_{\check{x},\check{y},\check{z}\leftrightarrow\hat{x},\hat{y},\hat{z}}}$);

13   $BP_{P_{\check{x},\check{y},\check{z}\leftrightarrow\hat{x},\hat{y},\hat{z}}} \leftarrow$ Beam.Material.Set(M_t); // M_t read from the MILP

          instance file

    end

  return $BP_{P_{\check{x},\check{y},\check{z}\leftrightarrow\hat{x},\hat{y},\hat{z}}}$

Ansys SpaceClaim: Volume Body Algorithm 5 models a structural member as a volume body, which includes two midpoints in the straight structural member segment, that is a part of a truss-like structure. The inputs to Algorithm 5 are the pre-processed material constitution M_t and structural member profile A_t (MILP instance file) of a structural member $B_{t,i,j}$ of type t (MILP solution file). In Line 9, the profile of the volume body is created. Note that in Lines 5 and 6, we get the magnitude $VM_{P_{\check{x},\check{y},\check{z}\leftrightarrow\hat{x},\hat{y},\hat{z}}}$ and direction $VD_{P_{\check{x},\check{y},\check{z}\leftrightarrow\hat{x},\hat{y},\hat{z}}}$ of the vector $V_{P_{\check{x},\check{y},\check{z}\leftrightarrow\hat{x},\hat{y},\hat{z}}}$ (see Line 4) to be able to create the frame $F_{P_{\check{x},\check{y},\check{z}}}$ (see Line 7) and the body $BO_{P_{\check{x},\check{y},\check{z}\leftrightarrow\hat{x},\hat{y},\hat{z}}}$ (see Line 10). Algorithm 5 returns a structural member as a volume body that is a part of a truss-like structure saved as a single top-level design file (*.scdoc).

Algorithm 5: ModelMemberVB, sketched

Algorithm ModelMemberVB (*solution_file, M_t, identifier*)

    input : MILP: *solution_file*, pre-processed material constitution M_t
            constructOR: *identifier*

    output: a structural member modeled as volume body from scratch; the structural member is an element of a truss-like structure as single top-level design file (*.scdoc)

    /* create all structural members as volume body; necessary structural members are read from the MILP solution file; pre-processed material constitution from the MILP instance file   */

1   **foreach** $B_{t,i,j} = 1$ *in solution_file* **do**

2       $P_{\check{x},\check{y},\check{z}} \leftarrow$ Point.Create($\check{x}, \check{y}, \check{z}$);

3       $P_{\hat{x},\hat{y},\hat{z}} \leftarrow$ Point.Create($\hat{x}, \hat{y}, \hat{z}$);

4       $V_{P_{\check{x},\check{y},\check{z}\leftrightarrow\hat{x},\hat{y},\hat{z}}} \leftarrow P_{\check{x},\check{y},\check{z}} - P_{\hat{x},\hat{y},\hat{z}}$;

5       $VM_{P_{\check{x},\check{y},\check{z}\leftrightarrow\hat{x},\hat{y},\hat{z}}} \leftarrow$ Vector.Magnitude.Get($V_{P_{\check{x},\check{y},\check{z}\leftrightarrow\hat{x},\hat{y},\hat{z}}}$);

6       $VD_{P_{\check{x},\check{y},\check{z}\leftrightarrow\hat{x},\hat{y},\hat{z}}} \leftarrow$ Vector.Direction.Get($V_{P_{\check{x},\check{y},\check{z}\leftrightarrow\hat{x},\hat{y},\hat{z}}}$);

7       $F_{P_{\check{x},\check{y},\check{z}}} \leftarrow$ Frame.Create($P_{\check{x},\check{y},\check{z}}, VD_{P_{\check{x},\check{y},\check{z}\leftrightarrow\hat{x},\hat{y},\hat{z}}}$);

8       $P_{P_{\check{x},\check{y},\check{z}}} \leftarrow$ Plane.Create($F_{P_{\check{x},\check{y},\check{z}}}$);

9       $PR_{P_{\check{x},\check{y},\check{z}}} \leftarrow$ CircleProfile.Create($P_{P_{\check{x},\check{y},\check{z}}}, A_t$); // A_t read from the MILP instance file

10      $BO_{P_{\check{x},\check{y},\check{z}\leftrightarrow\hat{x},\hat{y},\hat{z}}} \leftarrow$ Body.ExtrudeProfile($PR_{P_{\check{x},\check{y},\check{z}}}, VM_{P_{\check{x},\check{y},\check{z}\leftrightarrow\hat{x},\hat{y},\hat{z}}}$);

11      $DB_{P_{\check{x},\check{y},\check{z}\leftrightarrow\hat{x},\hat{y},\hat{z}}} \leftarrow$ DesignBody.Create(Document.MainPart, identifier, $BO_{P_{\check{x},\check{y},\check{z}\leftrightarrow\hat{x},\hat{y},\hat{z}}}$);

12      $DB_{P_{\check{x},\check{y},\check{z}\leftrightarrow\hat{x},\hat{y},\hat{z}}} \leftarrow$ DesignBody.Material.Set(M_t); // M_t read from the MILP instance file

    end

  return $DB_{P_{\check{x},\check{y},\check{z}\leftrightarrow\hat{x},\hat{y},\hat{z}}}$

Post-Processing

As can be seen in the schematic sketch in Figure 6.2 and the detail view in Figure 6.3, interferences from clashing bodies and unwanted free spaces inevitably occur due to the connection of the structural members with variable diameters at the intersection points. In accordance with Nguyen and Vignat (2016), in the case of solid, round structural members with a constant cross-section, it is necessary to correct the intersection points between the structural members.

Figure 6.2 Principle of the geometrical post-processing of interferences from clashing bodies: (left) truss-like structure without post-processing; (middle) interferences of the truss-like structure; (right) post-processed[7] truss-like structure

To post-process the intersection, a solid sphere needs to be added at each of the intersection points to fill the spaces (see Figure 6.2 right). The diameter of the added sphere's surface should be at least equal to the diameter of the structural member with the largest cross-section and, therefore, diameter. We model the sphere surface $S_{i,k,d}$ as a function of the work points $i \in V$, the maximum structural member diameter $d_{i,max}$ at work point $i \in V$, and a scaling factor $k \geq 1$, so that

$$S_{i,k,d} = k \cdot d_{i,max} \quad \forall i \in V \tag{6.1}$$

applies. For post-processing the overlapping structural members, they are segregated through splitting at their overlap limits (see the points a, b, c, d in Figure 6.3 left). Only one of the previously overlapping objects is retained so that the topology, which includes material distribution, is merged (see the post-processed truss-like structure in Figure 6.3 right). Post-processing excludes rough contour transitions. Furthermore, the self-weight of the connection nodes, which was falsified by material

---

[7] For a better illustration, Figure 6.2 right shows the sphere surfaces $S_{i,k,d}$ for a relatively large value of the scaling factor $k \geq 1$. In the course of the computational study of this thesis (see Chapter 7), we restrict ourselves to the case $k = 1$.

overlaps (cf. discussion starting in Section 5.1), is corrected. This post-processing makes the 3D-CAD data, including the external and self-weight loads, suitable for FEA and ready for AM. Stress concentrations and singularities (see, e.g., Wriggers 2008) are expected in the FEA for two reasons: the transitions of the sphere surfaces to the structural members and the transitions of multiple structural members with and without identical axes (e.g., point c in Figure 6.3 left). The fixing of the transition angle at $\beta = 90°$ (see Section 4.5) results in sharp re-entrant corners that can be triggers for stress concentrations and singularities. The post-processing of the intersection points is considered segregated in our 3D-CAD tool. For more information on a correct FEA and computational experiments, please refer to Section 7.3.

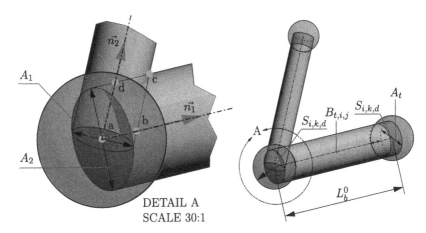

DETAIL A
SCALE 30:1

Figure 6.3 Detail view of the geometrical post-processing of interferences from clashing bodies: (left) the transitions of the sphere surfaces to the structural members and the transitions of multiple structural members; (right) $S_{i,k,d}$ as a function of the work points $i \in V$ and the the maximum structural member diameter $d_{i,max}$

Algorithm 6 adds a sphere surface at an intersection point if post-processing is necessary. The sphere surface gets merged with the existing bodies, i.e., interfering regions from clashing bodies are removed. The inputs to Algorithm 6 are the MILP instance and solution files and the profile type read from the library of standard beam profiles of Ansys SpaceClaim. In Line 1 of Algorithm 6, we search for the set CP of datum points shared by at least two structural members. If the set CP is not empty, we iterate over the set CP (see Line 2) and find the maximum cross-sectional area of a structural member using a datum point shared by at least two structural members. However, one must keep in mind that depending on the data structure used for storing all bodies in the top-level design and the bodies that have to be post-processed, three lists are generated in Lines 4, 5, and 14. Lines 6 and 7 are used to check whether any combinations of multiple structural members with an identical central axis have to be excluded. A sphere-shaped design body master (SpaceClaim Corporation 2019) is generated in Lines 8–13. Afterwards, in Line 18, the sphere surface is merged with all bodies saved in a list (see Lines 14–17). Finally, we merge all the remaining bodies available in the top-level design file (see Line 19). The post-processed truss-like structure is saved as a single top-level design file (*.scdoc).

Speed Up the Merge Process A very time-intensive part of our algorithm-driven product design process, as stated in Subsection 4.1.2, is the merging of the structural members and spheres of a truss-like structure. In order to counteract this, we have introduced a new two-part merging strategy in Algorithm 6. To emphasize the benefit of merging in multiple steps to speed up the merge process (see Lines 18 and 19 in Algorithm 6), we have compared the performance of just the standard merge function included in Ansys SpaceClaim 2020 R2 with that of our implementation in Table 6.1. The knowledge concerning the spheres that have to be merged with the overlapping structural members is incorporated into the modeling process of spheres (see Lines 4–5 and 14–18 in Algorithm 6). Regarding a 3D (2D) ground structure (see sets A_2, A_3 in Table 5.3), we split up the merge task into smaller tasks, with a maximum of 27 (9) bodies, which have to be merged directly after the modeling of one sphere. For speeding up the merge process, minimizing the number of bodies merged in one task is extremely effective and beneficial. A performance comparison of an increasing number of structural members modeled as volume bodies (first column) and the associated sphere bodies as volume bodies (second column) is given in Table 6.1. The third and fourth columns indicate the computational time needed by Algorithm 6 to generate the 3D-CAD data of an increasing number of spheres and merge the spheres with overlapping structural members, respectively. Column six indicates the computational time needed by the standard merge function included in Ansys SpaceClaim 2020 R2 to perform the same task. Columns five and seven indicate the overall computational time taken by each merging procedure, whereas columns eight and nine indicate the time difference between the two merging procedures. As shown in Table 6.1, utilizing Algorithm 6 resulted in a boost in both the number of merged structural members and spheres and the overall merging time. In generating 3000 structural members and 343 spheres, we were able to save 99.59% of the computation time by merging the truss-like structure using Algorithm 6. In addition, we were able to merge 50000 structural members and 4913 spheres instead of 3000 structural members and 343 spheres within the time limit of 75600 seconds.

Algorithm 6: PostProcessing, sketched

Algorithm PostProcessing(*instance_file, solution_file, profile_type, identifier*)

    input : $\underline{\text{MILP}}$: *instance_file, solution_file*

            construcTOR: *identifier; profile_type*

    output : post-processed truss-like structure

    `/* add a sphere surface at intersection point if necessary;`

    `   merge sphere with existing bodies; remove interfering regions`

    `   from clashing bodies                                        */`

1 $CP \leftarrow \bigcap\limits_{\substack{\{i,j\}\in B \\ B_{t,i,j}=1}} \{i,j\};$ `// find the set of datum points shared by at`

    `least two structural members in the` MILP `instance and solution`

    `file`

2   **foreach** $P \in CP$ *in Document.MainPart* **do**

3 $PR_{max} = \max\limits_{\substack{\text{Beams that intersects } DP \\ \forall t \in T, B_{t,i,j}=1}} \{profile\_type\}$

4 $BeamBodyList \leftarrow$ new List();

5 $BeamBodyList \leftarrow$ all $BO_{P_{\hat{x},\hat{y},\hat{z}\leftrightarrow\hat{x},\hat{y},\hat{z}}}$ in

      Document.MainPart.Bodies.Shape.Body;

6   **if** $B_{t,i,j} \neq B_{t,i,j}$; `//` $B_{t,i,j}$ `read from the` MILP `solution file`

      then

7   **if** $B_{t,i,j} \angle B_{t,i,j} \neq 180°$ **then**

8 $D \leftarrow$ Direction.Create(1, 1, 0);

9 $F_P \leftarrow$ Frame.Create(P, D);

10 $S_P \leftarrow$ Sphere.Create($F_P, PR_{max} \cdot S_{i,k,d}$);

11 $BoxUV \leftarrow new\ BoxUV();$

12 $BO_P \leftarrow$ Body.CreateSurfaceBody(S_P, BoxUV);

13 $DB_P \leftarrow$ DesignBody.Create(Document.MainPart, identifier,

         BO_P); `// Create a design body master`

14 $SphereBeamList \leftarrow$ new List();

15   **foreach** $BO_{P_{\hat{x},\hat{y},\hat{z}\leftrightarrow\hat{x},\hat{y},\hat{z}}}$ *in BeamBodyList* **do**

16   **if** $BO_{P_{\hat{x},\hat{y},\hat{z}\leftrightarrow\hat{x},\hat{y},\hat{z}}}.Contains(P)$ **then**

17 $SphereBeamList.\text{Add}(BO_{P_{\hat{x},\hat{y},\hat{z}\leftrightarrow\hat{x},\hat{y},\hat{z}}});$

           end

         end

        end

18 $BO_P.\text{Merge}(SphereBeamList);$

      end

    end

19   Document.MainPart.Bodies.Merge(); `// merge all remaining bodies`

    `using SpaceClaim`

    return *post-processed Ansys SpaceClaim design file (*.scdoc)*

Table 6.1 Comparison of the performance of Ansys SpaceClaim 2020 R2 with that of our SpaceClaim add-in constructOR in merging an increasing number of structural members modeled as volume bodies and associated sphere bodies. We solved instances within 75600 seconds

structural members[1] []	sphere bodies[2] []	modeling spheres [s]	own merging [s]	overall time [s]	default merge [s]	overall time [s]	time difference [s]	percentage difference [%]
500	125	1	11	12	120	121	109	90.08
1000	125	16	28	44	6960	6976	6932	99.37
2000	343	29	92	121	8100	8129	8008	98.51
3000	343	158	137	295	72660	72818	72523	99.59
4000	512	218	265	483	–	–	–	–
5000	512	487	304	791	–	–	–	–
10000	1331	1080	1543	2623	–	–	–	–
20000	2197	6000	5220	11220	–	–	–	–
40000	4096	29340	18780	48120	–	–	–	–
50000	4913	44580	27180	71760	–	–	–	–

[1] The calculations were performed on a workstation with an Intel Xeon E5-2637 v4 (3.5 GHz), 64 GB RAM and an NVIDIA GeForce RTX 2080 (8 GB RAM).
[2] Calculated using Ansys SpaceClaim

6.3　Performance Study

Autodesk Inventor Professional 2020

Using an Autodesk Inventor Professional add-in, i.e., Algorithms 1 and 3, we were able to generate 6084 structural members, each with a circular and constant cross-section, modeled as volume bodies in 20 hours, 46 minutes, and 30 seconds. The data processing of Autodesk Inventor Professional 2020 is based on the same BREP models for defining the topology and geometry of structural members as other 3D-CAD standard software are. Therefore, our results are representative of other 3D-CAD standard software such as Dassault Systems[8], CATIA V5, and PTC[30] Creo Elements/Direct. The performance of the Autodesk Inventor Professional add-in in designing a cubic truss-like structure was observed.

Increasing truss-like structure sizes, i.e., number of structural members (see Figure 6.4), were used to study the performance impact of an increasing number of structural members. The analysis of the computation time and memory usage of all intermediate file formats necessary in our algorithm-driven product design process (see Subsection 4.1.2) are presented in Table 6.2. In order to achieve an accurate evaluation according to industry practice, three different STL and G-code (standard RS-274) file quality levels were analyzed. A quadratic growth of the running time, which depends on the number of structural members to be designed, was observed. The Random Access Memory (RAM) used and the STEP (ISO 10303-242:2020-04)

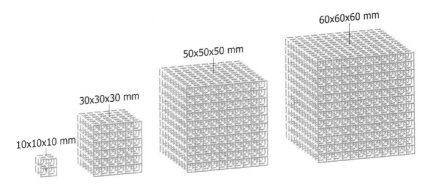

Figure 6.4　The truss-like structure sizes used to evaluate the computation time and memory requirements of our Autodesk Inventor Professional add-in

---

[8] Further information regarding the 3D-CAD software can be found on https://www.3ds.com/catia/ or https://www.ptc.com/creo/elements-direct/ (accessed February 28, 2021).

Table 6.2 Time and memory requirements for all the intermediate file formats used in our algorithm-driven product design process

$V\,[mm^3]$			10^3	30^3	50^3	60^3
CAD[1]	$\sum x_{i,j}$ [members]		54	882	3630	6084
	Time [s]		24	1755	24980	74790
	RAM [MB]		587	1850	5641	9150
	STEP[2] [MB]		0.24	3.99	17.02	63.61
CAE[1]	STL [MB] (ASCII/binary)	quality 1[3]	3/1	62/11	256/48	429/81
		quality 2[4]	7/1	123/23	508/96	849/161
		quality 3[5]	24/4	398/75	1637/311	2738/521
CAM[1]	G-Code [MB]	quality 1	0.30	4.65	19.17	31.92
		quality 2	0.32	5.01	20.82	34.17
		quality 3	0.34	5.04	22.31	31.19

[1] The calculations were performed on a workstation with an Intel Xeon E5-2637 v4 (3.5 GHz), 64 GB RAM and an NVIDIA GeForce RTX 2080 (8 GB RAM).
[2] Standard ISO 10303, AP242, spline fit accuracy 0.0001 mm, including sketches.
[3] Surface deviation 0.004%, normal deviation 30.00%, maximum edge length 100.00%, aspect ratio 21.50
[4] Surface deviation 0.016%, normal deviation 15.00%, maximum edge length 100.00%, aspect ratio 21.50
[5] Surface deviation 0.005%, normal deviation 10.00%, maximum edge length 100.00%, aspect ratio 21.50

file size were linear dependent on the number of structural members. The intermediate format STL was used in both the ASCII and the binary formats. Both intermediate file formats and the G-code required for CAM were linear dependent on the number of structural members.

Ansys SpaceClaim add-in construcTOR

We present the results of computational experiments on the generation of 3D-CAD data of an increasing number of circular and square structural members using our SpaceClaim add-in `construcTOR`. First, we studied the performance in the modeling of all structural members of a truss-like structure as volume bodies from scratch (see Algorithms 2 and 5 and Data Table 6.3). Second, we studied the performance in the modeling of all structural members of a truss-like structure as beam objects using the beam class of the Ansys SpaceClaim API (see Algorithms 2 and 4 and Data Table 6.3). In addition, we referred to the electronic material on pages 5–12, where all data sets (see Data Tables C.1–C.4) and more detailed computational experiments (see Figures C.1–C.5) are presented.

Volume Body Figures 6.5, C.1, and C.2 show that the computational time for generating circular and square structural members as volume bodies and the number of structural members were (almost) proportional to each other for all instance sizes. For up to 100000 structural members, i.e., smaller instances (see Figures C.1 and C.2 left), the computational time and the number of structural members were proportional to each other for all instance sizes. In addition, faceting the structural members (see, e.g., SpaceClaim Corporation 2019) caused an over-proportional increase in the computational time, which in turn resulted in an over-proportional increase in the overall computational time. The behavior of the faceting process, performed by Ansys SpaceClaim, was often irregular. This can particularly be seen in Figures 6.5 middle and C.2 right. Figure 6.5 right shows that modeling square structural members as volume bodies turned out to be faster than modeling square structural members for smaller instances.

In Figures 6.6, C.3, and C.4 are depicted the maximum memory (RAM) used during the entire process of generating 3D-CAD data of an increasing number of circular and square structural members modeled as volume bodies. It can be seen that the maximum memory consumption occured at the end of the faceting process. Figures 6.9 and 6.10 depict the memory usage over time in the modeling of circular and square structural members, respectively, as volume bodies. Memory consumption increased regularly during the generation and the faceting process of the structural members. At the end of the faceting process, a drastic increase in memory consumption was detected. We suspect that the reason for that was the visualization of the truss-like structure, started at the time by Ansys SpaceClaim. It is not intuitively clear why, in some cases, a horizontal pattern in the memory consumption existed. We suspect that Ansys SpaceClaim's strategies for cache and memory management were the reason. Figure 6.6 right shows that the maximum memory usage for generating circular and square structural members and the number of structural members were (almost) proportional to each other. In addition, the generation of circular structural members as volume bodies was less memory-intensive than modeling square structural members as volume bodies.

In summary, our Ansys SpaceClaim add-in `construcTOR` is able to model 400000 circular and square structural members as volume bodies within 3 hours, 12 minutes, and 38 seconds and 3 hours, 48 minutes, and 14 seconds, respectively (see Data Table 6.3). This shows that utilizing Algorithms 2 and 5 can massively speed up the modeling process and boost the number of modeled structural members. One should keep in mind that the results in the previous paragraph show that we were able to generate 6084 structural members with a circular and constant cross-section modeled as volume bodies in 20 hours, 46 minutes, and 30 seconds using Autodesk Inventor Professional 2020.

Beam Class Figure 6.7 shows that the computational time for generating circular and square structural members as beam objects using the beam class of the Ansys SpaceClaim API and the number of structural members were proportional to each other for all instance sizes. In addition, the faceting of the circular and square structural members caused an over-proportional increase in the computational time, which in turn resulted in an over-proportional increase in the overall computational time. The benefit of generating all beam profiles read from the MILP instance in a first loop (see Lines 1–8 in Algorithm 4) can be seen in Figure 6.7 right: The computational time is independent of the geometrical complexity of the structural member's cross-section. As a volume body does not have a beam profile as a data object, this approach cannot be implemented for generating structural members as volume bodies.

Figure 6.8 depicts the maximum memory used during the entire process of generating the 3D-CAD data of an increasing number of circular and square structural members modeled as beam objects. Figure 6.8 right shows that the generation of circular structural members as beam objects was less memory-intensive than modeling square structural members as beam objects. Figures 6.11 and 6.12 depict the memory usage over time in the modeling of circular and square structural members, respectively, as beam objects. It can be seen that the process of faceting was less memory-intensive than that of modeling structural members as volume bodies. Figure C.5 in the electronic supplementary material on page 11 demonstrates the performance of Algorithms 2 and 4 by drawing a comparison between the modeling of 15000 and 20000 circular as well as square structural members. In summary, our Ansys SpaceClaim add-in `construcTOR` is able to model 20000 circular and square structural members as volume bodies within 4 hours, 38 minutes, and 1 second and 4 hours, 44 minutes, and 56 seconds, respectively (see Data Table 6.3).

The results show that the process of modeling the structural members of a truss-like structure using the beam class of the Ansys SpaceClaim API is significantly slower than that of modeling the structural members as volume bodies, even for small instances. Nevertheless, it is crucial to note that a volume body and a beam object are different kinds of design data with differing characteristics (see, e.g., SpaceClaim Corporation 2014, 2019).

Table 6.3 The runtime and the maximum memory usage for generating increasing number of structural members depending on the type of implementation and the geometrical complexity of the structural member's cross section. The runtime and the maximum memory usage are divided into the generation of the 3D-CAD data of the structural members, the faceting of the structural members, and the overall runtime and memory usage. Ansys SpaceClaim, the beam class of the Ansys SpaceClaim API, our Ansys SpaceClaim add-in constructOR, and structural members modeled as volume bodies were used (see Algorithms 2 and 4)

type of implementation	structural members[1]	cross-section [mm²]	runtime [s]			maximum memory usage [MB]		
			generation	faceting	overall	generation	faceting	overall
beam class[3]	1000	circle	6	2	8	1024	1029	1030
	1000	square	6	2	8	1038	1047	1049
	5000	circle	34	202	236	1120	1123	1124
	5000	square	35	220	255	1138	1141	1142
	20000	circle	337	16344	16681	1352	1341	1352
	20000	square	389	16707	17096	1574	1499	1574
volume body[2]	20000	circle	81	232	313	1420	1929	2048
	20000	square	106	31	137	1819	2345	2438
	200000	circle	1113	2442	3555	4266	9423	9529
	200000	square	1395	3386	4781	8516	13660	13781
	400000	circle	2354	9204	11558	7404	17966	18048
	400000	square	2810	10884	13694	15714	25610	26677

[1] The calculations were performed on a workstation with an Intel Xeon E5-2637 v4 (3.5 GHz), 64 GB RAM and an NVIDIA GeForce RTX 2080 (8 GB RAM).
[2] Please refer to Data Tables C.3 and C.4 in the electronic supplementary material on pages 7 and 8 for the full data set.
[3] Please refer to Data Tables C.1 and C.2 in the electronic supplementary material on pages 5 and 6 for the full data set.

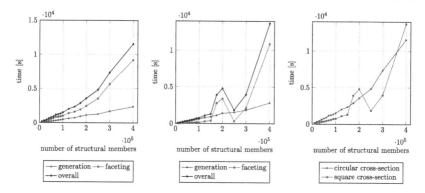

Figure 6.5 Computation time over an increasing number of circular and square structural members modeled as volume bodies using our Ansys SpaceClaim add-in construcTOR (see Algorithms 2 and 5 in Section 6.2 and Data Tables C.3 and C.4 in the electronic supplementary material on pages 7 and 8). The computation time is divided into the runtime for generating the 3D-CAD data of the structural members, the runtime for faceting the structural members, and the overall runtime: (left) circular structural members; (middle) square structural members; (right) overall computational time for circular and square structural members

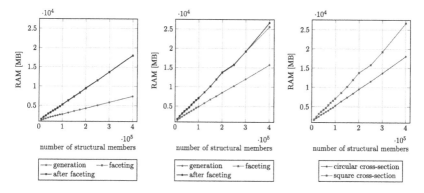

Figure 6.6 Maximum memory usage over an increasing number of circular and square structural members modeled as volume bodies using our Ansys SpaceClaim add-in construcTOR (see Algorithms 2 and 5 in Section 6.2 and Data Tables C.3 and C.4 in the electronic supplementary material on pages 7 and 8). The maximum memory usage is divided into the memory usage for generating the 3D-CAD data of the structural members, the memory usage for faceting the structural members, and the memory usage after faceting: (left) circular structural members; (middle) square structural members; (right) memory usage after faceting for circular and square structural members

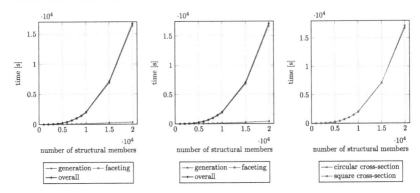

Figure 6.7 Computation time over an increasing number of circular and square structural members modeled as beam objects using the beam class of the Ansys SpaceClaim API and our Ansys SpaceClaim add-in construcTOR (see Algorithms 2 and 4 in Section 6.2 and Data Tables C.1 and C.2 in the electronic supplementary material on pages 5 and 6). The computation time is divided into the runtime for generating the 3D-CAD data of the structural members, the runtime for faceting the structural members, and the overall runtime: (left) circular structural members; (middle) square structural members; (right) overall computational time for circular and square structural members

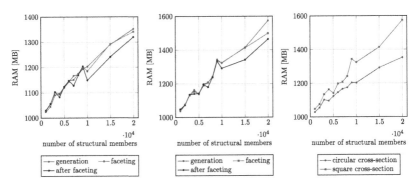

Figure 6.8 Maximum memory usage over an increasing number of circular and square structural members modeled as beam objects using the beam class of the Ansys SpaceClaim API and our Ansys SpaceClaim add-in construcTOR (see Algorithms 2 and 4 in Section 6.2 and Data Tables C.1 and C.2 in the electronic supplementary material on pages 5 and 6). The maximum memory usage is divided into the memory usage for generating the 3D-CAD data of the structural members, the memory usage for faceting the structural members, and the memory usage after faceting: (left) circular structural members; (middle) square structural members; (right) memory usage after faceting for circular and square structural members

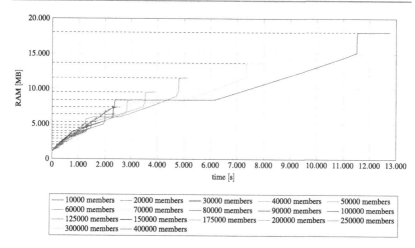

Figure 6.9 Memory usage over time for an increasing number of circular structural members modeled as volume bodies using our Ansys SpaceClaim add-in construcTOR (see Algorithms 2 and 5 in Section 6.2 and Data Tables C.3 and C.4 in the electronic supplementary material on pages 7 and 8)

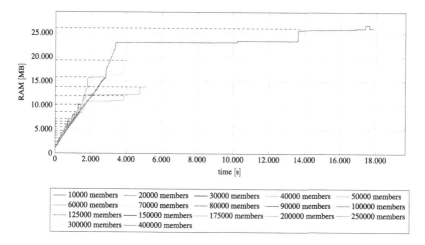

Figure 6.10 Memory usage over time for an increasing number of square structural members modeled as volume bodies using our Ansys SpaceClaim add-in construcTOR (see Algorithms 2 and 5 in Section 6.2 and Data Tables C.3 and C.4 in the electronic supplementary material on pages 7 and 8)

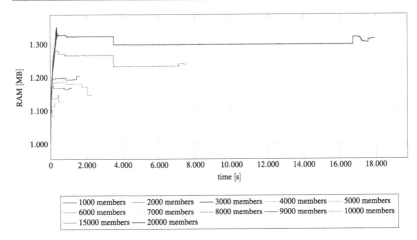

Figure 6.11 Memory usage over time for an increasing number of circular structural members modeled as beam objects using the beam class of the Ansys SpaceClaim API and our Ansys SpaceClaim add-in construcTOR (see Algorithms 2 and 4 in Section 6.2 and Data Tables C.1 and C.2 in the electronic supplementary material on pages 5 and 6)

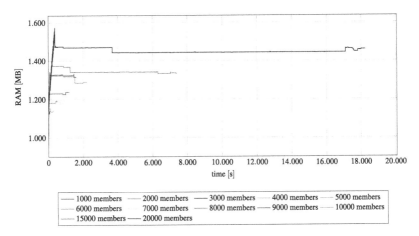

Figure 6.12 Memory usage over time for an increasing number of square structural members modeled as beam objects using the beam class of the Ansys SpaceClaim API and our Ansys SpaceClaim add-in construcTOR (see Algorithms 2 and 4 in Section 6.2 and Data Tables C.1 and C.2 in the electronic supplementary material on pages 5 and 6)

6.4   User Interface

To reduce the gap in knowledge between mathematical optimization and 3D-CAD, use our algorithm-driven product design process to its full potential, and make mathematical optimization of truss-like structures available to the 3D-CAD community, we provide a suitable User Interface (UI) in the Ansys SpaceClaim add-in construcTOR. The UI is designed for experienced 3D-CAD users familiar with the main design principles (see, e.g., Chang 2014, and the references therein) used in CAD and CAE. The first implementation draft of the UI was implemented in close cooperation with Emrah Dursun (Dursun 2019–2020). Hendrik Becker brought further improvements and performed further implementations within the scope of a bachelor's thesis (Becker 2020) under my supervision.

To bring consistency in both visual layout and user interactions with Ansys SpaceClaim, the add-in construcTOR has been designed according to the Ansys SpaceClaim add-in style guide (SpaceClaim Corporation 2014) and the Ansys SpaceClaim developer's guide (SpaceClaim Corporation 2019). The add-in is implemented in C#, which was developed on the .NET framework version 4.6.2[9]. The ribbon, i.e., the primary menu of the add-in (see Figure 6.13), has been designed

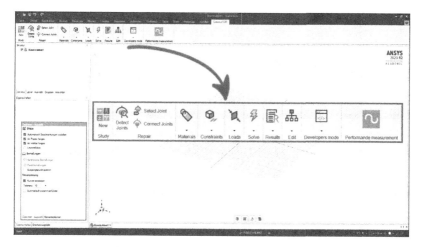

Figure 6.13 UI in the Ansys SpaceClaim add-in construcTOR: detail view of the ribbon

---

[9] Further information regarding .NET-based API from Microsoft can be found on https://docs. microsoft.com/en-us/dotnet/ (accessed February 28, 2021).

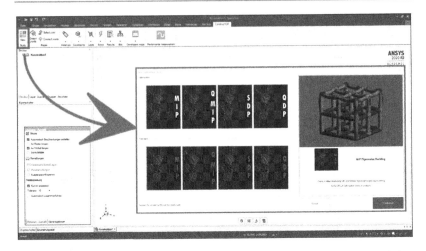

Figure 6.14 UI in the Ansys SpaceClaim add-in `constructTOR`: creating a new optimization study based on the mathematical optimization problem type

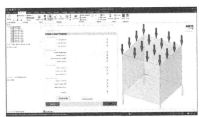

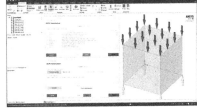

(a) Generate the mathematical optimization instance data from a 3D-CAD model of the design domain.

(b) Set the CPLEX starting information in the Ansys SpaceClaim environment.

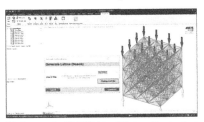

(c) Generate the 3D-CAD data of the truss-like structure from the CPLEX solution data.

(d) Post-process the 3D-CAD data of the truss-like structure automatically to obtain a ready-for-FEA and ready-for-machine-interpretation 3D-CAD model.

Figure 6.15 Workflow of the Ansys SpaceClaim add-in `constructTOR`

according to the 2007 Microsoft Office system UI design guidelines[10]. The ribbon has been designed modularly and from left to right. First, the user creates a new optimization study based on the mathematical optimization problem type that he wants to use in a later stage of the design optimization (see Figure 6.14). Next, the user generates the mathematical optimization instance data from a 3D-CAD model of the design domain (see Figure 6.15a), sets the CPLEX starting information in the Ansys SpaceClaim environment (see Figure 6.15b), generates the 3D-CAD data of the truss-like structure from the CPLEX solution data (see Figure 6.15c), and post-processes the 3D-CAD data of the truss-like structure automatically to obtain a ready-for-machine-interpretation 3D-CAD model (see Figure 6.15d).

---

[10] Further information regarding the 2007 Microsoft Office System UI Design Guidelines can be found on https://developer.microsoft.com/en-us/office/docs/ (accessed February 28, 2021).

Computational Study

7

To demonstrate the applicability and overall performance of our algorithm-driven product design process, we applied it to several design cases (instances) of spatial truss-inspired AM applications. The majority of the design cases in this chapter are based on accepted-for-publication or published conference papers and journals (see Section 1.3). We have accepted a certain degree of redundancy between the sections in order to gain the advantage of making all sections, i.e., design cases, mostly self-explanatory so that they can be read independently. Chapter 7 is divided into four sections. First, in Section 7.1, we have applied the MILP $TTO_{l;p}$, taking into consideration a 1940-member spatial truss cube with one static loading scenario manufactured as a functional prototype using SLS. In Section 7.2, we have applied the MILP $TTO_{l;s}$, taking into consideration a 1940-member spatial truss with one static loading scenario for designing self-supporting truss-like structures. In Section 7.3, we have applied the MILP $TTO_{l;m}$ to a demonstrator tool of a segmented blank holder, i.e., a 7325-member spatial truss with one static loading scenario. In Section 7.4, we have applied the QMIP $TTO_{l;q}$ to RTTO, taking into consideration a 296-member spatial truss tower with 128 static loading scenarios and a 1720-member spatial truss bridge with 8 static loading scenarios. The dimensions of the forces are in part not to scale in the figures and first angle orthographic projections (see, e.g., DIN ISO 128-34:2002-05) of this chapter. Note that the title block of the first angle orthographic projections' drawing sheets shows mathematical optimization data. The necessary material parameters have been defined in each design case using SI units. Some design cases are of purely academic nature since part of the values for the material parameters do not correspond to a physical material. At the end of

---

Supplementary Information The online version contains supplementary material available at https://doi.org/10.1007/978-3-658-36211-9_7.

C. Reintjes, *Algorithm-Driven Truss Topology Optimization for Additive Manufacturing*, https://doi.org/10.1007/978-3-658-36211-9_7

each design case, one optimized truss-like structure of the respective design case is presented in the form of a first angle orthographic projection. We already refer to the electronic supplementary material on pages 13–31 for further very insightful first angle orthographic projections of optimized truss-like structures for the Design Cases 7.3–7.5.

7.1 Optimization for Powder-Based Additive Manufacturing Systems: MILP TTO$_{I;p}$

Motivation
This truss-inspired design case investigates the application of our algorithm-driven product design process for AM. The procedures for automatically converting the mathematical optimization results generated by the MILP TTO$_{I;p}$ and CPLEX 12.6.1 (see Section 5.2) to a post-processed solid 3D-CAD model and for validation via FEA have been put into engineering practice. A functional prototype was manufactured using SLS to serve as a proof of concept.

Computational Results
The computational experiments for Design Case 7.1 and Design Case 7.2 were run on an Intel(R) Xeon(R) E5-2637 v3 with 3.60 GHz and 128 GB RAM using CPLEX Version 12.6.1 running on default but restricted to a single thread. Our Ansys SpaceClaim add-in constructTOR was implemented in Ansys SpaceClaim 2019 R3. The generation of 3D-CAD models of our truss-like structures inclusive geometry cleanup and simplification for FEA were done on a workstation with an Intel Xeon E5-2637 v4 with 3.5 GHz, 64 GB RAM, and an NVIDIA GeForce RTX 2080 with 8 GB RAM.

Design Case 7.1 (1940-member spatial truss cube with one static area load).
In the case of the 1940-member spatial truss cube with one static area load, we assumed a 100 mm × 100 mm × 100 mm reference volume $\mathbb{V}_1 \subseteq \mathbb{R}^3_+$ (see Subsection 4.4.2) with a 20 mm basic vertex distance; see Figures 7.1a and 7.1b. The reference volume $\mathbb{V}_1$ was modeled as a three-dimensional ground structure, with a set $V = \{1, \ldots, \overline{xyz}\} = \{1, \ldots, 6^3\} = \{1, \ldots, 216\}$ connection nodes and 1940 (calculated using Ansys SpaceClaim) potential structural members; see Figure 7.1b. A static area load at the last plane in the positive y-direction (top plane) was specified. The applied static area load $36 \cdot -3\,kN = -108\,kN$ was simplified by a purely vertical centric force application at each node of the top plane; see Figure 7.1b. The four

corner points of the first plane in the positive y-direction (bottom plane) were defined as the bearings; see Figure 5.4. We assumed two fixed bearings on the two lower-left corners and two floating bearings on the lower-right corners. The displacement at the bearings was fixed and the bearing capacity was considered to be infinite. A set of different pre-processed perfect structural members of types $T = \{0, 1, \ldots, 3\}$ with diameters $\{2, 4, 6, 8\}$ mm (see Subsection 4.3.1) together with the associated $c_{t,i,j} \in \{1067, 4270, 9611, 40000\}$ N (independent of i and j) and $cost_t \in \{28, 56, 84, 140\}$ € were specified as parameters. Note that these parameters do not correspond to any physical material. In addition, the reference volume $\mathbb{V}_1$ was given by the undeformed length of a structural member $L_b^0 \in \left\{20, 20\sqrt{2}, 20\sqrt{3}\right\}$ mm (see Equation (5.1) and Relation (5.2)) and the number of connection nodes in each direction in space $\overline{x}, \overline{y}, \overline{z} = 6$. For the sake of simplicity and abbreviation, please refer to Figures 7.1a–7.1c for the loading case $Q_i, i \in V$. We did not require the optimized truss-like structure to be symmetric with respect to a symmetry plane; compare the principle shown in Figure 5.4 with Figure 7.1c.

The computation time (runtime) using CPLEX 12.6.1 for the Design Case 7.1 (manually interrupted) was 10 h 51 min. Twelve permissible solutions were determined, whereby the optimality gap was 55.73%. The number of structural members could be reduced from 1940 (entire ground structure) to 665, which is a reduction of 65.72%. Please refer to Figure 7.3 for a first angle orthographic projection of the optimized truss-like structure based on the Design Case 7.1.

Numerical Analysis and the Additive Manufacturing Process
The optimized truss-like structure with a volume of $V = 58338$ mm^3 was manufactured with SLS; see Figures 7.1c and 7.2b. In contrast to the optimized truss-like structure, it was predetermined that the top plane was fully developed with beams with a diameter of 6 mm. The data preparation for the SLS manufacturing of the truss-like structure with the AM system EOS INT P770[1] was done with the rapid prototyping software Materialise Magics 3D Print Suite[2]. The Build Processor available in the Magics 3D Print Suite was used to generate the G-code file from the STL file. The STL file was generated with Ansys SpaceClaim. There was no post-processing of the STL data; the STL data obtained from our Ansys SpaceClaim add-in constructTOR was sufficiently high in resolution and the transitions were

---

[1] Further information regarding the technical data of the AM system can be found on https://www.eos.info/eos-p-770/ (accessed February 28, 2021).

[2] Further information regarding the data and build preparation software Materialise Magics 3D Print Suite can be found on https://www.materialise.com/en/ (accessed February 28, 2021).

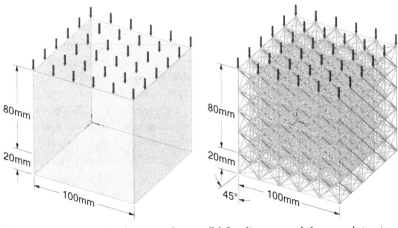

(a) Loading case and reference volume $\mathbb{V}_1 = \mathbb{V}_2$ for Design Cases 7.1 and 7.2

(b) Loading case and the ground structure for Design Cases 7.1 and 7.2

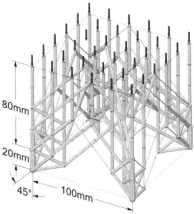

(c) Non-symmetric solution for Design Case 7.1: $D = \{2, 4, 6, 8\}$ mm, $V = 58338$ mm^3, optimality gap $= 55.73\%$

Figure 7.1 Design Case 7.1 and 7.2: (a) simplified loading case and reference volume $\mathbb{V}_1 = \mathbb{V}_2$; (b) loading case and the ground structure; (c) non-symmetric solution for Design Case 7.1, generated using the MILP TTO$_{\text{l;p}}$ and CPLEX 12.6.1

appropriately post-processed. To validate the mathematical optimization solution, we performed a linear elastic and nonlinear elastic numerical analysis with Ansys AIM 19.2[3]. It determined the deformation, stresses, and strains in our optimized truss-like structure as a function of the static loading case specified by the mathematical optimization instance data. Inertia loads and dead weight were analyzed, as they were simplified in the MILP TTO$_{l;p}$. Dynamic or damping effects were optional and were not considered in this validation. The boundary conditions for setting up the FEA were determined by the loading case and the bearings considered in the mathematical optimization instance data; see Figure 7.2a. The FEA mesh of the truss-like structure was a structured mesh without mesh refinements at the transitions of components, and tetrahedral elements were used. The element size was defined as 2.0 mm and the trial function was program-controlled. In Figure 7.2a, the total deformation has been provided as a fringe plot to aid in the visualization of the results of the linear elastic numerical analysis of the truss-like structure.

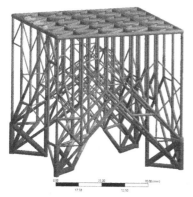

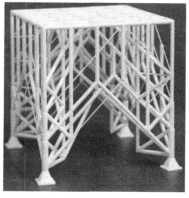

(a) Color-coded plot of the total deformation

(b) Additively manufactured truss-like structure with bearings as single components

Figure 7.2 Linear elastic numerical analysis of the solution for the Design Case 7.1, generated using the MILP TTO$_{l;p}$ and CPLEX 12.6.1, and the SLS manufactured functional prototype

---

[3] Further information on design engineering simulations with Ansys AIM can be found on https://www.ansys.com/ansys-aim-brochure.pdf (accessed February 28, 2021).

The truss-like structure was support-free, because the used AM system EOS INT P 770, as a plastic laser-sintering system for direct manufacturing, can create components without the need for support structures. No additional (physical) post-processing was necessary for the truss-like structure due to the design rules implemented in the MILPs $\text{TTO}_{l;p}$ and the post-processing in our Ansys SpaceClaim add-in constructOR. The material used for manufacturing was fine polyamide PA 2200 with a modified layer thickness of 0.12 mm.

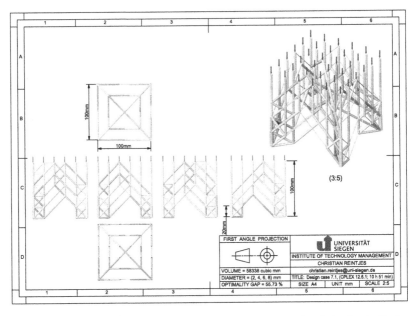

Figure 7.3 First angle orthographic projection of the non-symmetric solution for the Design Case 7.1: $D = \{2, 4, 6, 8\}$ mm; $V = 58338$ mm^3; optimality gap $= 55.73\%$. Solution generated using the MILP $\text{TTO}_{l;p}$ and CPLEX 12.6.1

7.2 Optimization of Support-Free Truss-Like Structures: MILP $\text{TTO}_{l;s}$

Motivation

Specific AM processes, e.g., FDM, possibly require internal or external, or both, support structures (see Subsection 2.2.3). Depending on the truss-like structure's geometric complexity, these support structures generate extra costs due to additional material, printing time, and energy. Furthermore, the force distribution in the truss-like structure is manipulated by the additional support structures, making any previous structural optimization useless; this is assuming that the support structures cannot be removed by non-destructive post-processing.

Instead of the optimization strategy of minimizing the need for additional support by optimizing the support structure itself while keeping the original truss-like structure, we focused on designing a new truss-like structure in a single step by optimizing the shape and topology for a support-free truss-like structure (see Section 4.5). Assuming that the support structures cannot be removed by non-destructive post-processing (see, e.g., the support structures in Figure 7.4, right)—which is very likely in the manufacturing of complex, lightweight truss-like structures with AM processes—this optimization approach is necessary since force distribution in a truss-like structure is manipulated by the additional support structures, which makes previous structural optimization invalid.

To solve this problem, the MILP TTO$_{I;s}$ (see Section 5.3), taking into consideration the linear design constraints for inclined and freestanding (support-free) structural members (cylinders) and the assumptions for location and orientation of structural members within a build volume of an AM system, has been presented. The aim was to realize support-free truss-like structures.

Computational Results

Design Case 7.2 (1940-member spatial truss cube with one static area load and self-supporting constraints).
For the case of the 1940-member spatial truss cube with one static area load and self-supporting constraints, we assumed a 100 mm × 100 mm × 100 mm reference volume $\mathbb{V}_2 = \mathbb{V}_1 \subseteq \mathbb{R}_+^3$ (see Subsection 4.4.2) with a 20 mm basic vertex distance; see Figures 7.1a and 7.1b. The reference volume $\mathbb{V}_2$ was modeled as a three-dimensional ground structure, with a set $V = \{1, \ldots, \overline{xyz}\} = \{1, \ldots, 6^3\} = \{1, \ldots, 216\}$ connection nodes and 1940 (calculated using Ansys SpaceClaim) potential structural members; see Figure 7.1b. A static area load at the last plane in the positive y-direction (top plane) was specified. The applied static area load $36 \cdot -3\ kN = -108\ kN$ was simplified by a purely vertical centric force application at each node of the top plane; see Figure 7.1b. The four corner points of the first plane in the positive y-direction (bottom plane) were defined as the bearings; see Figure 5.4. We assumed two fixed bearings on the two lower-left corners and two floating bearings on the lower-right corners. The displacement at the bearings was fixed and the bearing capacity was infinite. A set of different pre-processed perfect structural

members of types $T = \{0, 1, \ldots, 4\}$ with diameters $\{1, 2, 4, 6, 8\}$ mm (see Subsection 4.3.1) together with the associated $c_{t,i,j} \in \{1067, 4270, 9611, 40000\}$ N (independent of i and j) and cost$_t \in \{28, 56, 84, 140\}$ € were specified as parameters. Note that these parameters do not correspond to any physical material. In addition, the reference volume $\mathbb{V}_2$ was given by the undeformed length of a structural member $L_b^0 \in \left\{20, 20\sqrt{2}, 20\sqrt{3}\right\}$ mm (see Equation (5.1) and Relation (5.2)) and the number of connection nodes in each direction in space $\overline{x}, \overline{y}, \overline{z} = 6$. For the sake of simplicity and abbreviation, please refer to Figures 7.1a and 7.1b for the loading case $Q_i, i \in V$. We did not require the optimized truss-like structure to be symmetric with respect to a symmetry plane; compare the principle shown in Figure 5.4 with Figure 7.4 left.

Remark 7.1 (Connection between Design Cases 7.1 and 7.2).
The Design Case 7.2 for the MILP TTO$_{l;s}$ is identical to the Design Case 7.1 for the MILP TTO$_{l;p}$, except for the additional structural member with a diameter of 1 mm. We used both design cases to compare the solution for the MILP TTO$_{l;p}$ (without linear self-supporting constraints) with that of the MILP TTO$_{l;s}$ (with linear self-supporting constraints).

By assumption, the diameter of 1 mm is the minimal additively manufacturable diameter of a structural member. This assumption is based on the fact that the MILP TTO$_{l;s}$ should have the freedom to use this minimal structural member diameter to comply with the implemented linear design constraints for support-free truss-like structures while consuming minimum material. Since our geometry-based modeling in the MILP TTO$_{l;s}$, see Section 5.3, was coupled with the force equilibrium Constraints (5.11b) to (5.11d) and the moment equilibrium Constraints (5.11e) and (5.11f) via the binary variable $x_{i,j}$, there was no possibility of evaluating the resulting truss-like structure, regardless of the number of structural members used to avoid a combination of structural members at a node requiring a support structure. We considered the support structure in the preliminary stage of the design, i.e., during mathematical optimization, and therefore, as the entire force-bearing.

The computation time using CPLEX 12.6.1 for the Design Case 7.2 (manually interrupted) was 49 h 27 min. As many as 103 permissible solutions were determined, whereby the optimality gap was 64.46%.

We then statistically evaluated and compared the Design Case 7.1 for the MILP TTO$_{l;p}$ and the Design Case 7.2 for the MILP TTO$_{l;s}$. In order to allow comparability, we made the following assumptions for the Design Case 7.2: As AM process, FDM was performed on the AM system Ultimaker[4]s 2 Extended+. The support structure was designed with the supplied 3D printing software Ultimaker Cura 3.6.9. The support structure had been designed with a support overhang angle $\delta = 45°$, brim-type build platform adhesion, a grid pattern, 30% density, and free placement of support structures. The material was Acrylonitrile Butadiene Styrene (ABS), with a density of 1.1 g cm^{-3}. The layer thickness was set to 0.1 m; the members were printed as solid material.

Let us denote by $\mathbb{L}_2$ the weight of the truss-like structures generated by the MILPs TTO$_{l;p}$ and TTO$_{l;s}$ and CPLEX 12.6.1. As lightweight design criteria, we defined the ratio of $\mathbb{V}_1 = \mathbb{V}_2$ to $\mathbb{L}_2$. Table 7.1 shows that the model TTO$_{l;s}$ required $5.43 \cdot 10^4$ mm^3 less actual volume for the truss-like structure than the solution by Cura 3.6.9 did. This resulted in a 3.14% better ratio of $\mathbb{V}_2$ to $\mathbb{L}_2$ and a weight saving of 103.12 g, which corresponds to a cost saving of 46.10% in terms of material. The use of our MILP TTO$_{l;s}$ ensures that no support structure and post-processing is necessary. Note the difference between the solution for the MILP TTO$_{l;p}$ (see Figure 7.1c) and that of the MILP TTO$_{l;s}$ (see Figure 7.4 left).

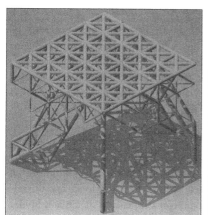 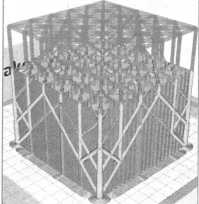

Figure 7.4 Influence of support structures: (left) support-free truss-like structure generated by the MILP TTD$_{l;s}$ and CPLEX 12.6.1; (right) the layer view of the optimized truss-like structure generated by the MILP TTO$_{l;p}$ and CPLEX 12.6.1, with grid pattern of the supporting structures generated with Cura 3.6.9; layer 700 of 1028

---

[4] Further information on the AM system Ultimaker 2 Extended+ and the software Ultimaker Cura can be found on https://ultimaker.com/software/ultimaker-cura/ (accessed February 28, 2021).

Table 7.1 Statistics for the solutions for the Design Cases 7.1 and 7.2 generated using the MILPs $TTO_{l;p}$ and $TTO_{l;s}$ and CPLEX 12.6.1, and the solution generated by Cura 3.6.9. The second column shows the number of structural members, independent of their specific diameter. Column four shows the ratio of $\mathbb{V}_2$ to $\mathbb{L}_2$. The fifth and sixth columns stand for the weight of the truss-like structure and the optimality gap, respectively

model	$\sum x_{i,j}$[1]	$\mathbb{V}_2$	$\mathbb{L}_2$	ratio	weight	gap
	[ ]	$[10^4 mm^3]$	$[10^4 mm^3]$	[%]	[g]	[%]
$TTO_{l;p}$	528	172.80	5.42	3.14	103.40	55.73[2]
$TTO_{l;s}$	875	172.80	6.36	3.68	121.00	64.46[3]
Cura 3.6.9	528	172.80	11.79	6.82	224.12	–

[1] Number of structural members after optimization.
[2] Computation time 10 h 51 min; 12 permissible solutions.
[3] Computation time 49 h 27 min; 103 permissible solutions.

7.3   Segmented Blank Holder: MILP $TTO_{l;m}$

Motivation

In tool-bound forming processes, e.g., car body drawing, the static geometry of the active surfaces is responsible for the geometry of the manufactured component. As shown in Figure 7.5, process improvement for avoiding wrinkling (see, e.g., Borchmann et al. 2020) or compensating for springback (see, e.g., Frohn-Sörensen et al. 2020) must be iteratively done through separate contour changes of the solid tool. According to Reuter (2020) the use of flexible and intelligent tools (see, e.g., Kuhnhen et al. 2020) will help avoid additional surface modifications and stabilize forming processes in a closed loop control. These tools require a suitable truss-like structure for force transmission and actuators below the surface.

Therefore, in this section we have presented an algorithm-driven optimization design study based on our MILP $TTO_{l;m}$ (see Section 5.4). The loads and boundary conditions have been taken from the numerical simulations of a blank holder (Kuhnhen et al. 2020); see Figures 7.6 and 7.7 for illustration. Taking into consideration lightweight construction and topology optimization, we have applied our algorithm-driven optimization workflow to additively manufactured forming tools inspired by Frohn-Sörensen et al. (2020). We have performed mathematical optimization, numerical shape optimization, and verification via numerical simulation.

Based on the algorithm-driven optimization workflow, we have optimized a demonstrator tool of a segmented blank holder (Kuhnhen et al. 2020). Finally, in this section we have offered an outlook on how an optimized truss-like structure can be used as a mechanism for in-process modification of local surface geometry and local structural stiffness (see, e.g., Kuhnhen et al. 2020, and the references therein).

Some of the ideas presented as part of Design Case 7.3 resulted from intensive discussions with Reuter (2020). The work was done in close cooperation with the Institute of Production Technology, Chair of Forming Technology[5], at the University of Siegen.

Tool-bound Forming Technology
Following Reuter (2020), the forming tools play a key role as the link between semi-finished products and machines and directly impact the flexibility of a forming process (Cao et al. 2019). State-of-the-art forming tools (see, e.g., Groth et al. 2018, Sörensen et al. 2020, and the references therein) are typically solid and oversized steel components, leading to an unnecessarily high level of energy consumption in tool production along the entire value chain and in the operation of the tools. This research gap can be addressed by combining lightweight construction with topology optimization to obtain an efficient design tool for forming tool development. On account of the fact that AM methods enable the fabrication of complex-shaped and topology-optimized tools (Cao et al. 2019)—in comparison to conventional manufacturing methods—the combination of lightweight construction, topology optimization, and AM is of significant interest.

Besides the established and in industrial finite element software, implemented continuum (stress-constrained) topology optimization methods based on SIMP (Saadlaoui et al. 2017), algorithm-driven optimization based on mathematical programming (Reintjes et al. 2019) can also be used for early-stage design optimization of truss-like structures. Reintjes and Lorenz (Reintjes and Lorenz 2020) showed a large-scale TTO of additively manufactured truss-like structures based on the high performance of (heuristic-based) optimization algorithms implemented in commercial linear programming software such as CPLEX.

---

[5] Further imformation regarding the University Siegen's Chair of Forming Technology can be found on https://protech.mb.uni-siegen.de/uts/en/ (accessed February 28, 2021).

MILP TTO$_{l;m}$ and its Application to a Segmented Blank Holder
Inspired by the works Kuhnhen et al. (2020), Frohn-Sörensen et al. (2020) and
Borchmann et al. (2020) we investigated sensoric and actuatoric forming tools with
the aim of designing self-adjustable surfaces; see Figure 7.5. Following Reuter
(2020), possibly, future forming tools will have the self-adjusting capability to control material flow and react to changing process conditions. A simple demonstrator for such a flexible tool has been shown in Figure 7.6. The segmented blank
holder consists of a housing with thread holes at the bottom, a cover, and a segmented inlay structure for force transmission; see the front view of the section
illustrated in Figure 7.7. The surface adjustment can be realized by the infeed of
one screw per segment. Experimental testing and numerical simulations were carried out with different arrangements and infeeds of the screws (Kuhnhen et al.
2020). The basic proof of concept was obtained by measuring the surface deformation using Gom ARAMIS Professional[6], which showed different surface profiles
depending on the screw setup (Kuhnhen et al. 2020). We examined how such a
force-transmitting inlay could be designed using (symmetric) truss-like structures
generated by the MILP TTO$_{l;m}$ and CPLEX 12.6.1. First, a linear static FEA using
Altair OptiStruct 14.0.210[7] was performed to obtain the load spectrum (instance
parameter) for the MILP TTO$_{l;m}$. We assumed that the insert was loaded by a screw
force of $F_{screw} = 4.5$ kN and a contact pressure between workpiece and inlay,
resulting in the process force $F_{process} = 13.5$ kN; see Figures 7.7 and 7.8a. The
reaction load is the contact pressure p_{cover} between the cover and the inlay. After a
transformation of the stress given in the FEA (linear elastic material behavior given
by nonlinear constraints) into linear and optimization-friendly constraints (centric
point loads applied to nodes), we got a formulation suitable for our MILP TTO$_{l;m}$.
To avoid numerical difficulties while computing a solution with CPLEX 12.6.1, we
rounded up the data of the stress given in the FEA to a slightly more accurate data.

---

[6] Further information regarding high-precision deformation analysis can be found on
https://www.gom.com/metrology-systems/aramis/ (accessed February 28, 2021).
[7] Further information regarding the generative design (topology optimization) tool can be
found on https://www.altair.com/structures-applications/ (accessed February 28, 2021).

flexibility level 1
contact pressure
distribution

flexibility level 2
springback
compensation

flexibility level 3
adaption of
tool contour

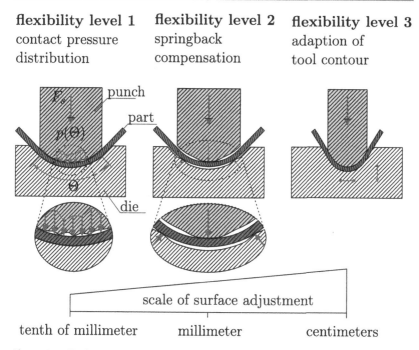

Figure 7.5 Flexibility levels of forming tools depending on the scale of surface adjustment (Kuhnhen et al. 2020)

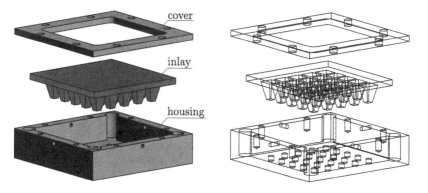

Figure 7.6 The additively manufactured demonstrator of a segmented blank holder. Note that the surface adjustment is realized by the infeed of screws using the fittings integrated into the housing (Kuhnhen et al. 2020)

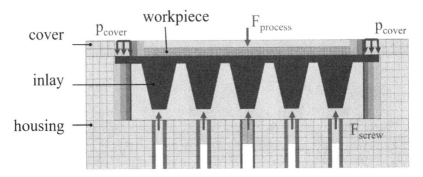

Figure 7.7 Sectional view of the demonstrator of a segmented blank holder to show the loading case for the FEA (Reuter 2020)

Computational Results

The computational experiments were run on an Intel(R) Xeon(R) E5-2637 v3 with 3.60 GHz and 128 GB RAM using CPLEX Version 12.6.1 running on default but restricted to a single thread. Our Ansys SpaceClaim add-in `constructOR` was implemented in Ansys SpaceClaim 2020 R2. The generation of the 3D-CAD models of our truss-like structures inclusive geometry cleanup and the simplification for FEA were performed on a workstation with an Intel Xeon E5-2637 v4 with 3.5 GHz, 64 GB RAM, and an NVIDIA GeForce RTX 2080 with 8 GB RAM.

Design Case 7.3 (7325-member spatial truss with one static loading scenario). *In the case of the segmented blank holder, i.e., a 7325-member spatial truss with one static loading scenario, we assumed a 100 mm × 100 mm × 50 mm reference volume $\mathbb{V}_3 \subseteq \mathbb{R}^3_+$ (see Subsection 4.4.2) with a 10 mm basic vertex distance; see Figures 7.8a and 7.8b. The process force $F_{process} = 81 \cdot -167\,N = -13527\,N$ (inner force) and the reaction force $F_{cover} = 28 \cdot -965.25\,N = -27027\,N$ (outer force) were defined at the last plane in the positive y-direction (top plane); see Figure 7.8a. The first plane in the positive y-direction (bottom plane) was loaded by the screw force $F_{screw} = 9 \cdot +4506 = +40554\,N$ realized by a specific infeed of the nine screws. The reference volume $\mathbb{V}_3$ was modeled as a three-dimensional ground structure, with a set $V = \{1, \ldots, 11 \cdot 6 \cdot 11\} = \{1, \ldots, 726\}$ connection nodes and 7325 (calculated using Ansys SpaceClaim) potential structural members; see Subsection 4.4.3 and Figure 7.8b. No bearings existed. The force F_{screw} acted as counterforce to the force $F_{process}$ and F_{cover}. The displacement at the nodes where F_{screw} applied* *was fixed. The considered material was aluminum AlSi10MG/3.2381[8] (EN AW-43000), with a yield strength of $\sigma_y = 190 \pm 0\ N\,mm^{-2}$ after stress-relief heat treatment and an elastic modulus of $E = 70 \pm 0\ GPa$. The yield strength and elastic modulus were assumed to be constant along the circular and uniform cross-section of a structural member. Pre-processed structural members with lower ($A_{\min} \in \{0.79, 3.14, 7.07\}\ mm^2$) and upper ($A_{\max} = 78.54\ mm^2$) bounds on the cross-sectional areas of the structural members together with a factor of safety $S = 1$ were specified as parameters; see Section 5.4 and Table 7.2. The reference volume $\mathbb{V}_3$ was given by $L_e = L_b^0$ and the length $L_e \in \left\{10, 10\sqrt{2}, 10\sqrt{3}\right\}$ mm of the edge e (see Equation (5.1) and Relation (5.2)) and the set of edges E (see Table 5.5). For the sake of simplicity and abbreviation, please refer to Figure 7.8a for all external forces **F**. We required the optimized truss-like structures to be symmetric with respect to two symmetry planes; compare the principle shown in Figure 5.4 and Figures 7.8a, 7.8c–7.8e.*

We examined the optimization results for several values of $A_{\min}$, but have referred to the minimum diameter $D_{\min}$ for the purpose of presentation. As for the case of the segmented blank holder, an overview of the optimized truss-like structures for $D_{min} = \{1, 2, 3\}$ mm have been shown in Figures 7.8c–7.8e. Figure 7.12 shows a first angle orthographic projection of the optimized truss-like structure based on the Design Case 7.3 for $D_{min} = 1$ mm. For the first angle orthographic projections of the optimized truss-like structures for $D_{min} = \{2, 3\}$ mm, please refer to the electronic supplementary material on pages 14–15. Table 7.2 displays the computational results and the limit of the runtime for each design case. The table indicates that for increasing $D_{\min}$, obtaining small optimality gaps is computationally more expensive, partially due to the worsening of the corresponding LP-relaxation. An increase in $D_{\min}$ leads to a decrease in the number of structural members.

Finite Element Analysis and Shape Optimization
To validate the mathematical optimization results, linear static FEAs were performed using Altair OptiStruct 14.0.210. The loading case was analogous to the loading case shown in Figure 7.8a. The truss-like structures shown in Figures 7.8c–7.8e were discretized with solid elements of type CTETRA with a nominal element edge length of 0.5 mm. Note that through this volumetric mesh, each node of the truss-like structure could transmit rotary moments, which is contrary to the assumptions of the MILP TTO$_{l;m}$. Another difference between the MILP TTO$_{l;m}$ (i.e., all our MILPs

---

[8] The material datasheet can be found on https://www.prototec.de/ (accessed February 28, 2021).

and the QMIP) and the FEA was in the material behavior: While the MILP $TTO_{l;m}$ could not consider the constitutive material equations without costly linearization, a linear elastic material (MATL1) was implemented in the FEA model with an elastic modulus of aluminum of $E = 70$ GPa.

Table 7.2 Computational results for the Design Case 7.3 for $D_{min} = \{1, 2, 3\}$ mm and $D_{max} = 10$ mm

D_{min} [mm]	D_{max} [mm]	best found [mm^3]	bound [mm^3]	gap [%]	runtime [s]	first found time [s]	first found value [mm^3]
1[1]	10	22815	22714	0.44	969828	7193	23756
2[2]	10	33622	23822	29.15	1032300	2029	49233
3[2]	10	56377	27192	51.77	362779	3214	86809

[1] For a first angle orthographic projection, please refer to Figure 7.12.
[2] For a first angle orthographic projection, please refer to the electronic supplementary material on pages 14–15.

The results of the FEAs have been shown in Figure 7.9, wherein, for simplicity, we have taken advantage of the two-fold symmetry and visualized just a quarter of the model. We found that the stresses in all three models were, in general, below the yield strength of $\sigma_y = 0.19$ GPa (see Design Case 7.3). From this, we conclude that the design suggested through mathematical optimization is a solution with good mechanical performance and geometrical properties for this static loading case. Nevertheless, it was found that some higher-stressed positions exist.

To overcome this problem, we suggest adding an finite element shape optimization to our algorithm-driven product design process, as described in Figure 4.3. To this end, we identified high-stressed areas (value of the actual stress greater than the yield strength of $\sigma_y = 0.19$ GPa) whose shape OptiStruct is allowed to change with a minmax objective function, to minimize the stress response while maximizing the material. This finite element shape optimization has been shown in Figure 7.10 for one high-stressed connection node of the optimized truss-like structure based on the Design Case 7.3 for $D_{min} = 1$ mm. In the initial state, there was a maximum stress of about 0.5 GPa. After 3 iterations of shape optimization, the surface was somewhat modified, and after 15 iterations we saw the final state, in which the upper bound stress constraint of 0.19 GPa was satisfied (see Figure 7.11 for the principle of a forming tool with in-process adjustable active tool surfaces).

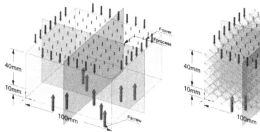

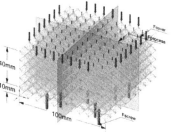

(a) The loading case derived by considering the FEA and reference volume $\mathbb{V}_3$ for Design Case 7.3

(b) The loading case and the ground structure for Design Case 7.3

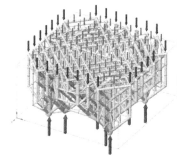

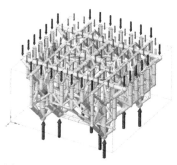

(c) The symmetric solution for Design Case 7.3: $D_{min} = 1$ mm; $V = 22815$ mm³; optimality gap = 0.44%

(d) The symmetric solution for Design Case 7.3: $D_{min} = 2$ mm; $V = 33622$ mm³; optimality gap = 29.15%

(e) The symmetric solution for Design Case 7.3: $D_{min} = 3$ mm; $V = 56377$ mm³; optimality gap = 51.77%

Figure 7.8 Design Case 7.3: (a) the simplified loading case and reference volume $\mathbb{V}_3$; (b) the loading case and the ground structure; (c–e) symmetric solutions for $D_{min} = \{1, 2, 3\}$ mm, generated using the MILP TTO$_{l;m}$ and CPLEX 12.6.1. (This is a joint work with Reuter 2020; based on Reintjes et al. 2021)

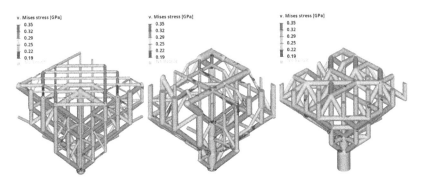

Figure 7.9 Comparison of the FEA, i.e., the color-coded von Mises stress distribution above the yield strength of $\sigma_y = 0.19$ GPa, and of the three optimized truss-like structures of Design Case 7.3: (left) $D_{min} = 1$ mm; (middle) $D_{min} = 2$ mm; (right) $D_{min} = 3$ mm. (This is a joint work with Reuter 2020; based on Reintjes et al. 2021)

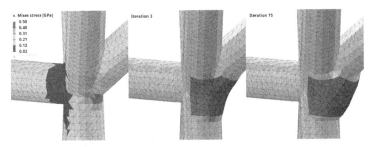

Figure 7.10 Shape optimization of a high-stressed connection node: (left) von Mises stress distribution in the initial state; (middle) free-shape design region of the connection node after 5 optimization iterations; (right) free-shape design region of the connection node after 15 optimization iterations. (This is a joint work with Reuter 2020; based on Reintjes et al. 2021)

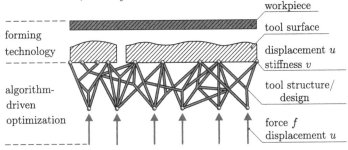

Figure 7.11 The principle of a forming tool with in-process adjustable active tool surfaces. Note that the tool structure includes technical joints. (This is a joint work with Reuter 2020; based on Reintjes et al. 2021)

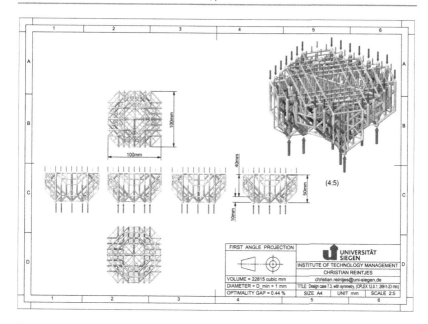

Figure 7.12 First angle orthographic projection of the symmetric solution for the Design Case 7.3: $D_{min} = 1$ mm; $V = 22815$ mm^3; optimality gap $= 0.44\%$. Solution generated using the MILP TTO$_{l;m}$ and CPLEX 12.6.1

7.4   Robust Truss Optimization: QMIP TTO$_{l;q}$

Motivation

Worst-case analysis of truss-like structures is essential in engineering for protection against failure in all practical situations encountered during a truss-like structure's lifetime; it is also required by safety regulation laws. Instead of determining and optimizing the worst-case scenario based on engineering experience, which is a process prone to human error, it seems more promising to optimize all meaningful loading scenarios using QMIP, without needing to identify the worst-case scenario (see, e.g., Ederer et al. 2011, Hartisch et al. 2016, Lorenz and Wolf 2015, Wolf 2015, s). For more information on this topic please refer to Hartisch (2020), Hartisch and Lorenz (2019a,2019b, 2020) and the references therein. We have considered robust TTO with multiple loading scenarios. Therefore, we have presented here an algorithm-driven optimization design study based on our QMIP TTO$_{l;q}$ (see Section 5.5). A typical dimensioning method is to identify and examine a suspected worst-case scenario using experience and component-specific information and also to incorporate a factor of safety to hedge against uncertainty. In this section, we have presented the QMIP TTO$_{I;q}$, which allowed us to specify expected scenarios without having explicit knowledge about the worst-case scenarios as the resulting optimal truss-like structure can withstand all the specified scenarios individually. This leads to less human error and higher efficiency and, thus, to the saving of both time and cost in a (algorithm-driven) product design process. Additionally, having a trustworthy initial design allows for the reduction of the factor of safety, which leads to material and cost saving. As part of the Design Cases 7.4 and 7.5, we have presented spatial trusses with minimal volume, which are stable for up to 100 loading scenarios. Additionally, the effect of demanding a symmetric truss-like structure and explicitly limiting the diameter of the truss members in the model on the computational costs has been discussed. This chapter is a joint work with Michael Hartisch (Hartisch 2016–2020).

Computational Results

For each QMIP instance, we built a corresponding DEP (Bertsimas et al. 2011). The computational experiments for Design Cases 7.4 and 7.5 were run on an Intel(R) Core(TM) i7-4790 with 3.60 GHz and 32 GB RAM using CPLEX Version 12.6.1 running on default but restricted to a single thread. Our Ansys SpaceClaim add-in construcTOR was implemented in Ansys SpaceClaim 2020 R2. The generation of optimized 3D-CAD models of our truss-like structures inclusive geometry cleanup and simplification for FEA were performed on a workstation with an Intel Xeon E5-2637 v4 with 3.5 GHz, 64 GB RAM, and an NVIDIA GeForce RTX 2080 with 8 GB RAM. We kept the runtime within the limit of 240 hours.

296-member spatial truss tower with 128 static loading scenarios

Design Case 7.4 (296-member spatial truss tower with 128 static loading scenarios).

In the case of the 296-member spatial truss tower with 128 static loading scenarios, we assumed a 20 mm × 40 mm × 20 mm reference volume $\mathbb{V}_4 \subseteq \mathbb{R}^3_+$ (see Subsection 4.4.2) with a 10 mm basic vertex distance; see Figures 7.13–7.17. We have assumed $C = 7$ loading cases and are interested in truss-like structures that can withstand each of the $2^C = 2^7$ static loading scenarios resulting from combining individual loading cases, as discussed in Section 5.5. For the sake of simplicity and abbreviation, see Figure 7.14a for all color-coded loading cases. The reference volume $\mathbb{V}_4$ was modeled as a three-dimensional ground structure, with a set $V = \{1, \ldots, 3 \cdot 5 \cdot 3\} = \{1, \ldots, 45\}$ connection nodes and 296 (calculated using Ansys SpaceClaim) potential structural members; see Subsection 4.4.3. The four *corner points of the first plane in the positive y-direction (bottom plane) were defined as the bearings; see Figure 5.4. We assumed two fixed bearings on the two lower-left corners and two floating bearings on the lower-right corners. The displacement at the bearings was fixed and the bearing capacity was considered to be infinite. The claimed material was fine polyamide PA 2200 Top Speed 1.0[9], with a yield strength of $\sigma_y = 45 \pm 0$ N mm^{-2} (as-build) and an elastic modulus of $E = 1.5 \pm 0$ GPa. The yield strength and elastic modulus were assumed to be constant along the circular and uniform cross-section of a structural member. Pre-processed structural members with lower ($A_{min} \in \{0, 0.79, 3.14, 7.07, 12.57, A_{VDI}\}$ mm^2) and upper ($A_{max} = 78.54$ mm^2) bounds on the cross-sectional areas of the structural members together with a factor of safety $S = 1$ were specified as parameters; see Section 5.5 and Table 7.3. The reference volume $\mathbb{V}_4$ was given by $L_e = L_b^0$ and the length $L_e \in \left\{ 10, 10\sqrt{2}, 10\sqrt{3} \right\}$ mm of the edge e (see Equation (5.1) and Relation (5.2)) and the set of edges E (see Table 5.7). Here, we have discussed the optimized truss-like structures with and without two-fold symmetry. The implementation of the symmetry was analogous to Design Case 7.3; see Figures 7.8a and 7.8b for illustration.*

The individual loading cases for $D_{min} = \{0, 1\}$ mm have been shown in Figures 7.13a and 7.14a. Additionally, optimal truss-like structures for each individual case and $D_{min} = \{0, 1\}$ mm have been displayed in Figures 7.13b–7.13h and 7.14b–7.14h to ensure better comprehensibility. The solution shown in Figure 7.13b gains in interest if we realize that it contains the same amount of structural members (not volume) as the solution shown in Figure 7.14b. Notwithstanding both solutions show totally different characteristics in terms of buckling (see Subsection 2.1.3). Accepting the possibility of a physically impractical solution for Design Case 7.4, the QMIP TTO$_{l;q}$ was allowed to build truss-like structures not connected to the bearings, i.e., we have used the existence of bearing reaction forces of zero. For this reason, it is of interest to look at the optimal but physically impractical solutions shown in Figures 7.13c and 7.14c. From Figures 7.13h and 7.14h, the drawback of not demanding additional zero-force-structural members (see Subsection 2.1.3 and Figure 2.9) is quite apparent: Local instability of the four long structural members could occur when the compression force exceeds its critical buckling load, such that multiple unstable states of equilibrium (no trivial state of equilibrium) can arise.

Please refer to Figure 7.18 for a first angle orthographic projection of the optimal solution for the Design Case 7.4 without symmetry and $D_{min} = 0$ mm. For first angle

---

[9] The material datasheet can be found on https://eos.materialdatacenter.com/ (accessed February 28, 2021).

orthographic projections of the non-symmetric solutions for $D_{min} = \{1, 3\}$ mm and symmetric solutions for $D_{min} = \{0, 1, 2, 3\}$ mm, please refer to the electronic supplementary material on pages 16–21. In Figure 7.15, a compilation of the best found robust solutions for $D_{min} = \{0, 1, 3\}$ mm without symmetry and $D_{min} = \{0, 2, 3\}$ mm with symmetry has been displayed; all solutions were stable in each of the $2^C = 2^7$ loading scenarios. Table 7.3 contains the objective values of the best found solutions, the best lower bounds, and the corresponding optimality gaps for different settings.

For $D_{min} = 0$ mm, the DEP of the QMIP instance could be pre-processed to be an LP problem as the binary $\mathbf{x}$ variables can be fixed to 1. The corresponding optimal solutions were found to be within 2 and 0.5 hours for the non-symmetric and symmetric cases, respectively. For the remaining instances, except for the symmetric instance with $D_{min} = 1$ mm, the optimality gap was not closed sufficiently within 240 hours. Table 7.3 indicates that with increasing D_{min}, obtaining small optimality gaps becomes computationally more expensive, which is partially due to the worsening of the corresponding LP-relaxation.

The solutions differed considerably: An increase in D_{min} led to a decrease in the number of structural members, and when the VDI design rules (see Equation (5.9) and (5.10)) were additionally enforced, long members—particularly diagonal members—were avoided; see Figure 7.16. Notably, only the truss-like structure given in Figure 7.15a seemed to be optimal. From Figure 7.15d, the drawback of demanding symmetry is quite apparent: The sufficient triangular structure at the bottom is no longer feasible. For the first angle orthographic projections of the (non-)symmetric solutions for $D_{min} \equiv D_{VDI}$, please refer to the electronic supplementary material on pages 22–23.

Table 7.3 Computational results for the Design Case 7.4, with and without demanding structural symmetry

D_{min} [mm]	without symmetry			with symmetry		
	best found [mm^3]	bound [mm^3]	gap [%]	best found [mm^3]	bound [mm^3]	gap [%]
0	3693	3693	0	4367	4367	0
1	3783	3697	2.28	4400	4399	0.01
2	4987	3771	24.37	5246	4674	10.90
3	7793	4286	45.00	8679	5860	32.48
4	13008	5330	59.02	12931	8045	37.79
D_{VDI}	6317[1]	3954	37.40	5906	5218	11.65

[1] For this instance, we continued the computation for 30 days, achieving a best found solution of 5267mm^3, barely improved bound, and a gap of 24.77%.

Symmetry   Demanding structural symmetry reduced the number of continuous **a**
and binary **x** variables from 296 to 94 as, only for the representative edges, these
decisions had to be made; see Figure 5.4. This resulted in the achievement of a
computational benefit, which has been reflected in Table 7.3, particularly in the
comparison of the optimality gaps. Obviously, the volume of the *optimal* symmetric
truss-like structure cannot be lower than the one without symmetry. Nevertheless, in
some cases the incumbent symmetric solution had lower volume upon reaching the
time limit. Since a primary goal in engineering is to reduce the (algorithm-driven)
product development cycle time, this fact should be encountered in practical TTO
and RTTO. The solutions for $D_{min} = 0$ mm and $D_{min} = 1$ mm reflect the price
of demanding a symmetric truss-like structure: For this instance, the volume of the
symmetric solution increased by about $\frac{1}{6}$.

Minimal diameter   The manufacturable minimal diameter depends on the manufacturing process and is obviously always larger than zero. Although the computational results for $D_{min} = 0$ mm invited the use of these quickly obtained solutions, the resulting truss-like structures exhibited numerous structural members with
extremely small diameters of < 0.1 mm, which needed to be removed to ensure
manufacturability. Hence, solving this LP formulation is computationally efficient
but unsuitable for direct application in the field of AM. However, in order to utilize this stable solution, one can inflate small members until the desired diameter
is attained; see Figure 7.17. Table 7.4 shows the resulting volumes of the truss-
like structures when—starting from the optimal but not manufacturable solution for
$D_{min} = 0$ mm—the diameter of the affected members was increased to the actual
value of D_{min}. In all cases, the volume dramatically exceeded the corresponding best
truss-like structure found during the optimization process; cf. Table 7.3. Therefore,
when disregarding the runtime, solving the model with explicitly stated minimum
diameter is preferred over post-processing the quickly obtained optimal solutions for
$D_{min} = 0$ mm. For the first angle orthographic projections of the (non-)symmetric
solutions for D_{min} inflated to D_{VDI} and $D_{min} = 1$ mm, please refer to the electronic
supplementary material on pages 24–25.

Table 7.4 Truss volumes [mm^3] of the optimized truss-like structures of the Design Case
7.4, with and without symmetry, when inflating members to D_{min} based on the LP solution.
Note that the second column shows the LP solution

D_{min}	0 mm (LP)	1 mm	2 mm	3 mm	4 mm	VDI
without symmetry	3693	4443	9208	18883	33106	16262
with symmetry	4367	4963	9522	19599	34319	17503

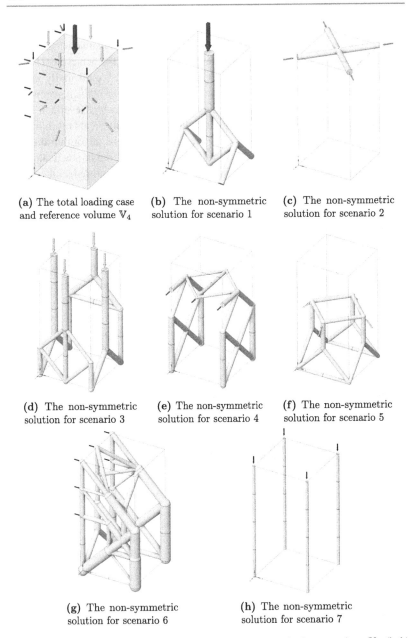

(a) The total loading case and reference volume $\mathbb{V}_4$

(b) The non-symmetric solution for scenario 1

(c) The non-symmetric solution for scenario 2

(d) The non-symmetric solution for scenario 3

(e) The non-symmetric solution for scenario 4

(f) The non-symmetric solution for scenario 5

(g) The non-symmetric solution for scenario 6

(h) The non-symmetric solution for scenario 7

Figure 7.13 Design Case 7.4: (a) the simplified loading case and reference volume $\mathbb{V}_4$; (b–h) the non-symmetric ($R(e) = e$) single loading cases for $D_{\min} = 0$ mm. (Based on Hartisch et al. 2021)

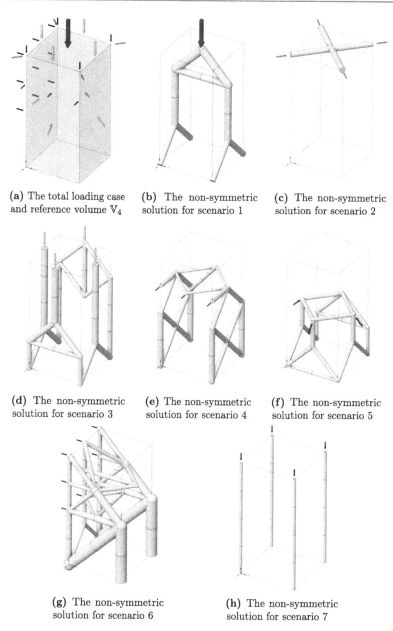

(a) The total loading case and reference volume $\mathbb{V}_4$

(b) The non-symmetric solution for scenario 1

(c) The non-symmetric solution for scenario 2

(d) The non-symmetric solution for scenario 3

(e) The non-symmetric solution for scenario 4

(f) The non-symmetric solution for scenario 5

(g) The non-symmetric solution for scenario 6

(h) The non-symmetric solution for scenario 7

Figure 7.14 Design Case 7.4: (a) the simplified loading case and reference volume $\mathbb{V}_4$; (b–h) the non-symmetric ($R(e) = e$) single loading cases for $D_{\min} = 1$ mm. (Based on Hartisch et al. 2021)

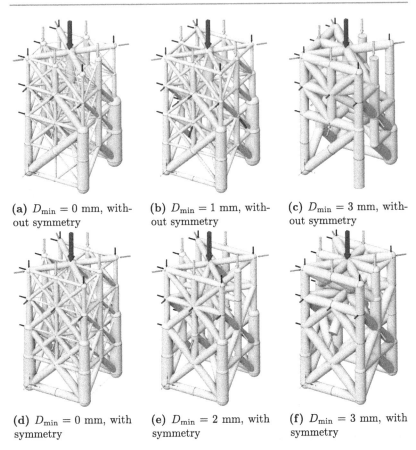

(a) $D_{\min} = 0$ mm, without symmetry

(b) $D_{\min} = 1$ mm, without symmetry

(c) $D_{\min} = 3$ mm, without symmetry

(d) $D_{\min} = 0$ mm, with symmetry

(e) $D_{\min} = 2$ mm, with symmetry

(f) $D_{\min} = 3$ mm, with symmetry

Figure 7.15 Design Case 7.4: (a–c) non-symmetric ($R(e) = e$) solutions for $D_{\min} = \{1, 2, 3\}$ mm; (d–f) symmetric solutions for $D_{\min} = \{1, 2, 3\}$ mm. (Based on Hartisch et al. 2021)

1720-member spatial truss bridge with 8 static loading scenarios

Design Case 7.5 (1720-member spatial truss bridge with 8 static loading scenarios).
In the case of the 1720-member spatial truss bridge with 8 static loading cases, we assumed a 90 mm × 30 mm × 40 mm reference volume $\mathbb{V}_5 \subseteq \mathbb{R}_+^3$ (see Subsection 4.4.2) with a 10 mm basic vertex distance; see Figure 7.19. Figure 7.19a shows

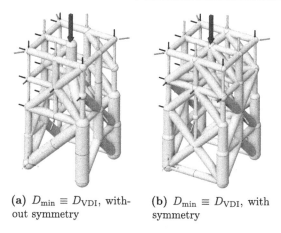

(a) $D_{min} \equiv D_{VDI}$, without symmetry

(b) $D_{min} \equiv D_{VDI}$, with symmetry

Figure 7.16 Design Case 7.4: (a) non-symmetric solution for $D_{min} \equiv D_{VDI}$; (b) symmetric solution for $D_{min} \equiv D_{VDI}$. (Based on Hartisch et al. 2021)

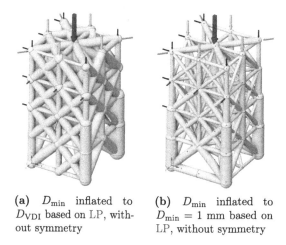

(a) D_{min} inflated to D_{VDI} based on LP, without symmetry

(b) D_{min} inflated to $D_{min} = 1$ mm based on LP, without symmetry

Figure 7.17 Design Case 7.4: (a) non-symmetric solution when inflating members to $D_{min} \equiv D_{VDI}$ based on the LP solution; (b) non-symmetric solution when inflating members to $D_{min} = 1$ mm. (Based on Hartisch et al. 2021)

all color-coded static loading cases. We have assumed 8 loading cases and are interested in truss-like structures that can withstand each individual loading case, as discussed Section 5.5. For the sake of simplicity and abbreviation, see Figure 7.19a

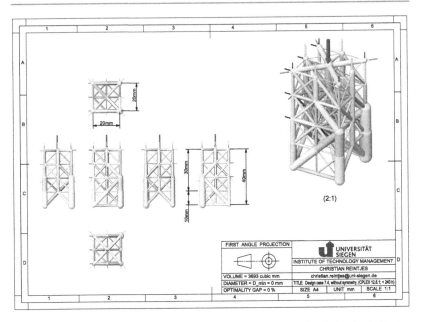

Figure 7.18 First angle orthographic projection of the non-symmetric solution for the Design Case 7.4: $D_{min} = 0$ mm; $V = 3693$ mm^3; optimality gap $= 0.00\%$. Solution generated using the QMIP TTO$_{l;q}$ and CPLEX 12.6.1

for the total color-coded loading case. The reference volume $\mathbb{V}_5$ was modeled as a three-dimensional ground structure, with a set $V = \{1, \ldots, 10\cdot4\cdot5\} = \{1, \ldots, 200\}$ connection nodes and 1720 potential structural members (calculated using Ansys SpaceClaim); see Subsection 4.4.3. The four corner points of the first plane in the positive y-direction (bottom plane) were defined as the bearings; see Figure 5.4. We assumed two fixed bearings on the two lower-left corners and two floating bearings on the lower-right corners. The displacement at the bearings was fixed and the bearing capacity was considered to be infinite. The claimed material was fine polyamide PA 2200 Top Speed 1.0[10] with a yield strength of $\sigma_y = 45 \pm 0$ N mm^{-2} (as-build) and an elastic modulus of $E = 1.5\pm0$ GPa. The yield strength and elastic modulus were assumed to be constant along the circular and uniform cross-section of a structural member. Pre-processed structural members with lower ($A_{min} \in \{0, 0.79, 3.14, 7.07, 12.57, A_{VDI}\}$ mm^2) and upper ($A_{max} = 78.54$ mm^2) bounds

---

[10] The material datasheet can be found on https://eos.materialdatacenter.com/ (accessed February 28, 2021).

on the cross-sectional areas of the structural members together with a factor of
safety S = 1 were specified as parameters (see Section 5.5). The reference volume
$\mathbb{V}_5$ *was given by* $L_e = L_b^0$ *and the length* $L_e \in \left\{ 10, 10\sqrt{2}, 10\sqrt{3} \right\}$ *mm of the edge*
e (see Equation (5.1) and Relation (5.2)) and the set of edges E (see Table 5.7).
We have discussed the optimized truss-like structures with two-fold symmetry. The
implementation of the symmetry was analogous to Design Case 7.3; see Figures
7.8a and 7.8b for illustration.

In Table 7.5, the computational results and the amount of structural members for
various values of D_{min} have been presented. Only for $D_{min} = 0$ mm (see the
electronic supplementary material on page 26) was the optimal truss-like structure
found (in 104 minutes). In Figure 7.19, the 8 color-coded loading cases as well as the
best found solutions for $D_{min} = \{0, 1, 2, 3\}$ mm and $D_{min} \equiv D_{VDI}$ for three different
optimality gaps, respectively, have been displayed. For comparison, the optimal
truss-like structure for $D_{min} = 0$ mm exhibited 1004 structural members, and for
$D_{min} \equiv D_{VDI}$ the not optimal truss-like structure exhibited 320 structural members.
Hence, changing the minimal diameter significantly alters the truss-like structure's
topology, regardless of its volume. For the first angle orthographic projections of the
symmetric truss-like structures for $D_{min} = \{0, 1, 2, 3\}$ mm and $D_{min} \equiv D_{VDI}$ for
two different optimality gaps, please refer to the electronic supplementary material
on pages 26–31. Please refer to Figure 7.20 for a first angle orthographic projection of
the optimized truss-like structure for $D_{min} \equiv D_{VDI}$ and an optimality gap = 50.19%.

Table 7.5 Computational results for the 1720-member spatial truss bridge with 8 static loading scenarios

D_{min}	0 mm[1]	1 mm[1]	2 mm[1]	3 mm[1]	4 mm[1]	VDI$_1^2$	VDI$_2^2$	VDI$_3^2$
best found solution [mm³]	6436	7096	16964	35124	61133	15318	16208	19202
best lower bound [mm³]	6436	6445	7435	11027	17606	7630	7629	7629
optimality gap [%]	0.00	9.18	56.17	68.60	71.20	50.19	52.93	60.27
number of structural members	1004	479	414	389	382	320	350	383

[1] For a first angle orthographic projection, please refer to the electronic supplementary material
on pages 26–31.
[2] To show the fundamental importance of the optimality (gap) in TTO, we analyzed three
different optimality gaps; see Figure 7.20 and the electronic supplementary material on pages
30–31.

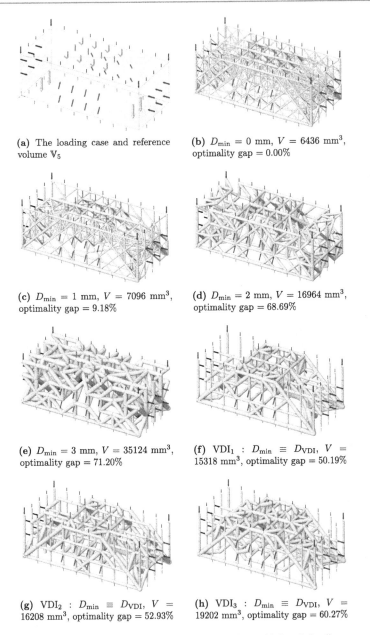

(a) The loading case and reference volume $\mathbb{V}_5$

(b) $D_{min} = 0$ mm, $V = 6436$ mm^3, optimality gap $= 0.00\%$

(c) $D_{min} = 1$ mm, $V = 7096$ mm^3, optimality gap $= 9.18\%$

(d) $D_{min} = 2$ mm, $V = 16964$ mm^3, optimality gap $= 68.69\%$

(e) $D_{min} = 3$ mm, $V = 35124$ mm^3, optimality gap $= 71.20\%$

(f) VDI$_1$: $D_{min} \equiv D_{VDI}$, $V = 15318$ mm^3, optimality gap $= 50.19\%$

(g) VDI$_2$: $D_{min} \equiv D_{VDI}$, $V = 16208$ mm^3, optimality gap $= 52.93\%$

(h) VDI$_3$: $D_{min} \equiv D_{VDI}$, $V = 19202$ mm^3, optimality gap $= 60.27\%$

Figure 7.19 Design Case 7.5: best found symmetric solutions with 8 static loading scenarios stable in each individual loading case. (Based on Hartisch et al. 2021)

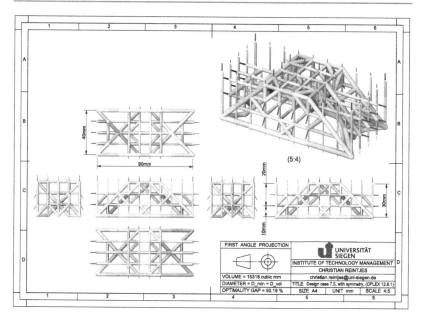

Figure 7.20 First angle orthographic projection of the non-symmetric solution for the Design Case 7.5: $D_{\min} \equiv D_{\mathrm{VDI}}$; $V = 15318\mathrm{mm}^3$; optimality gap $= 50.19\%$. Solution generated using the QMIP TTO$_{l;q}$ and CPLEX 12.6.1

Conclusion and Outlook

8

8.1 Conclusion

In this thesis, we presented how to systematically combine linear optimization, CAD, finite element shape optimization, FEA, and AM into an algorithm-driven product design process for additively manufactured truss-like structures. The main contributions of this thesis are threefold.

First, with the development of the MILPs $TTO_{l;p}$, $TTO_{l;s}$, $TTO_{l;m}$, and the QMIP $TTO_{l;q}$, which allow (quantified) linear optimization focusing on a) design rules, limitations, and standards for AM technologies, b) support-free truss-like structures, and c) loading uncertainty, we make global optimization of (quantified) MIPs, i.e., the opportunity to find the global solution for a TTO problem with the performance of state-of-the-art mathematical solvers, available to the structural optimization and AM community.

Second, by providing linear formulations of minimum weight TTO problems adapted for AM, initially implying a high degree of nonlinearity in the physics of the problems, we move linear optimization closer to real-world structural optimization in lightweight design for AM.

Third, with the development and implementation of our Ansys SpaceClaim add-in constructOR, we enable the generation of reliable 3D-CAD models of truss-like structures out of raw mathematical optimization data ready-for-machine-interpretation. Post-processing of intersections, geometry cleanup, and simplification for FEA are implemented. We reduce the gap in knowledge between linear optimization and CAD and make linear optimization of truss-like structures available to the CAD community.

Our MILP $TTO_{l;p}$ and the solver CPLEX was used to optimize a 1940-member spatial truss-like structure suitable for powder-based AM processes discounting

© The Author(s), under exclusive license to Springer Fachmedien Wiesbaden GmbH, part of Springer Nature 2022
C. Reintjes, *Algorithm-Driven Truss Topology Optimization for Additive Manufacturing*, https://doi.org/10.1007/978-3-658-36211-9_8

support structures. The optimized truss-like structure serves as a functional prototype (proof of concept) and was manufactured using the powder-based AM system
EOS INT P770. Our second MILP $TTO_{1;s}$ and the solver CPLEX was used to optimize a 1940-member spatial self-supporting truss-like structure, by considering
geometry-based design constraints for inclined and support-free structural members and assumptions for the location and orientation of structural members within
a build volume of an AM system.

We further applied our algorithm-driven product design process to the real-world
application of designing an additively manufactured lightweight forming tool, using
the example of a flexible blank holder. As a minimum cross-sectional area is essential due to design restrictions in AM and as symmetry can be exploited to effectively optimize structural systems, we used our MILP $TTO_{1;m}$, considering minimal
cross-sectional areas of the structural members depending on AM limitations and
standards, and a design-variable linking technique to enforce two-fold symmetry.
Contrary to the MILPs $TTO_{1;p}$ and $TTO_{1;s}$ we modeled continuous cross-sectional
areas. Our Ansys SpaceClaim add-in construcTOR was used for pre- and post-
processing the CPLEX solutions. Using finite element shape optimizations, highly
stressed areas were geometrically modified, resulting in an improved and overall
usable design. Finite element-based simulations were performed to validate preliminary designs conducted by the preceding mathematical optimization.

In addition, we utilized the QMIP $TTO_{1;q}$ to introduce a robust formulation of
our MILP $TTO_{1;m}$. Instead of determining and optimizing a worst-case scenario,
which is in accordance with engineering experience and prone to human error,
our approach allowed to state loading cases while ensuring the resulting truss-
like structure is stable, even in the (unknown) worst-case scenario. A distinction
was made about whether the loading cases only occurred individually or in any
sort of combination. We have presented results on a 296-member spatial truss-like
structure by considering the combination of seven loading cases, which resulted in
128 loading scenarios. In addition, a 1720-member spatial truss-like structure with
eight individually occurring loading scenarios was considered.

In a detailed computational study, we evaluated the applicability of our algorithm-
driven product design process for additively manufactured truss-like structures.
Additionally, we highlighted the advantages and disadvantages of explicitly enforcing a minimal cross-sectional area of structural members, AM standards, and a
symmetric structure in our models. We demonstrated using real-world design cases
that our algorithm-driven product design process is an efficient and reliable optimization tool for preliminary designs of truss-like structures. In addition, our Ansys
SpaceClaim add-in construcTOR is capable of obtaining well-performing and
ready-for-machine-interpretation CAD data out of raw mathematical optimization

data. In general, the research output of this thesis is the first step toward making the structural optimization, AM, and mathematical optimization community cooperate.

8.2   Outlook

The present work provides the foundation for an algorithm-driven product design process for additively manufactured truss-like structures. The promising computational results and the additively manufacturing of a functional prototype indicate that the algorithm-driven product design process can be successfully applied to real-world structural optimization in lightweight design for AM. In particular, minimum weight truss design problems with structural member stress constraints adapted for AM can be modeled as MILP or QMIP and used for CAD-based mathematical optimization.

However, our mathematical optimization solutions were computed using the standard solver CPLEX. To solve large-scale MILPs and QMIPs of practical relevance, it is likely that external heuristics, especially starting solutions in the solution strategy, and the development of more targeted lower and upper bounds can lead to a significant performance improvement both in time and memory usage when solving our TTO for AM problems. Furthermore, incorporating more AM limitations and standards could lead to an increase in component quality. In addition, future work must focus on the distortion of the objective value due to overlapping structural members at the connection nodes and the modeling of a penalization of the mass value of a connection node via the objective function. Also, the mathematical optimization models will gain interest if we implement unit cells with known mechanical behavior to solve even bigger instances.

To move linear optimization closer to real-world structural optimization in lightweight design for AM, implementing process-specific geometrical limitations of AM technologies will be of interest. Delamination of layers, curling, or stair-step effects should be minimized by formulating linear constraints for maximizing the component quality during mathematical optimization. If a support structure is necessary, minimizing its material should also be considered. The upshot of this is the possibility that the load-bearing truss-like structure and the support structure could be optimized separately by taking post-processing effort of AM techniques into consideration.

In addition, there is still a need for discussion since the degrees of freedom of a connection node in the FEA differ from the degrees of freedom in our mathematical optimization models. We claim that a connection node in our mathematical optimization models cannot transmit rotary moments. On the contrary, due to the post-processing of the optimized truss-like structures into merged volumes and the consequent volumetric meshing, a connection node in the FEA can transmit rotary moments. This is one of the reasons for the stress peaks in the FEA.

Another reason is that no constitutive material equations and no geometry have been implemented in our mathematical optimization models, since they can only contain linear constraints by nature. Therefore, the mathematical optimization models cannot take local stresses into account, which underlines the importance of our algorithm-driven product design process containing finite element shape optimizations to redesign highly stressed areas and finite element-based simulations for final validation.

Further work on our Ansys SpaceClaim add-in construcTOR, would help us to establish a component library, including technical joints and variable-length structural members. Based on the analysis of the interaction between local tool surface properties of the flexible blank holder and the forming result, we could define process-time dependent, necessary displacement, and stiffness at the links between the force-transmitting truss-like structure and forming tool surface. A new method based on our algorithm-driven product design process will be investigated to fulfill these requirements. We propose to build mechanisms for adjustable surfaces and structural stiffness through technical joints instead of a solid volume at a connection node or at variable-length structural members.

Apart from the improvement in algorithm-driven TTO for AM, we hope that the algorithm-driven product design process for additively manufactured truss-like structures proposed in this thesis finds an audience within the structural optimization, CAD, and AM community, insofar as optimal designs early in the design process are generated and costs due to design failure are minimized. Maybe within a few years, mathematical optimization plays such an essential role in the field of structural optimization in lightweight design for AM as topology optimization of continuum structures nowadays.

Bibliography

W. Achtziger and M. Kočvara. Structural topology optimization with eigenvalues. *SIAM Journal on Optimization*, 18(4):1129–1164, 2008.

W. Achtziger and M. Stolpe. Truss topology optimization with discrete design variables— guaranteed global optimality and benchmark examples. *Structural and Multidisciplinary Optimization*, 34 (1):1–20, 2007.

W. Achtziger and M. Stolpe. Global optimization of truss topology with discrete bar areas— part I: Theory of relaxed problems. *Computational Optimization and Applications*, 40(2): 247–280, 2008.

W. Achtziger and M. Stolpe. Global optimization of truss topology with discrete bar areas— part II: Implementation and numerical results. *Computational Optimization and Applications*, 44 (2):315–341, 2009.

W. Achtziger, M. P. Bendsøe, A. Ben-Tal, and J. Zowe. Equivalent displacement based formulations for maximum strength truss topology design. *IMPACT of Computing in Science and Engineering*, 4 (4): 315–345, 1992.

S.-H. Ahn, M. Montero, D. Odell, S. Roundy, and P. K. Wright. Anisotropic material properties of fused deposition modeling ABS. *Rapid Prototyping Journal*, 8 (2): 248–257, 2002.

A. Ahrari and K. Deb. An improved fully stressed design evolution strategy for layout optimization of truss structures. *Computers & Structures*, 164: 127–144, 2016.

S. Allen and D. Dutta. On the computation of part orientation using support structures in layered manufacturing. In *1994 International Solid Freeform Fabrication Symposium*, 1994.

L. Altherr, T. Ederer, U. Lorenz, P. Pelz, and P. Pöttgen. Experimental validation of an enhanced system synthesis approach. *Operations Research Proceedings 2014: Selected Papers of the Annual International Conference of the German Operations Research Society (GOR)*, pages 1–7, 2016.

L. C. Altherr. *Algorithmic System Design under Consideration of Dynamic Processes*. Shaker Verlag, 2016.

American Institute of Steel Construction. Manual of steel construction: allowable stress design. *American Institute of Steel Construction (AISC), Chicago, IL*, 1989.

ANSYS. Ansys SpaceClaim: 3D Modeling Software. https://www.ansys.com/de-de/products/3d-design/ansys-spaceclaim, 2019. Accessed February 2021.

J. Arora, M. Huang, and C. Hsieh. Methods for optimization of nonlinear problems with discrete variables: a review. *Structural Optimization*, 8 (2–3): 69–85, 1994.

© The Editor(s) (if applicable) and The Author(s), under exclusive license to
Springer Fachmedien Wiesbaden GmbH, part of Springer Nature 2022
C. Reintjes, *Algorithm-Driven Truss Topology Optimization for Additive Manufacturing*, https://doi.org/10.1007/978-3-658-36211-9

S. Arora and B. Barak. *Computational Complexity: a Modern Approach*. Cambridge University Press, 2009.

ASCENT. *Autodesk Inventor 2020: ILogic (Mixed Units)*. ASCENT—Center for Technical Knowledge, 2019.

ASCE/SEI 10–15. Design of latticed steel transmission structures. Standard, *American Society of Civil Engineers*, 2015.

ASTM F2792-12a. Standard terminology for additive manufacturing technologies (withdrawn 2015). Standard, *International Organization for Standardization, ASTM International*, West Conshohocken, PA, Sept. 2012.

C. Atwood, M. Griffith, L. Harwell, E. Schlienger, M. Ensz, J. Smugeresky, T. Romero, D. Greene, and D. Reckaway. Laser engineered net shaping (lens^{TM}): A tool for direct fabrication of metal parts. In *International Congress on Applications of Lasers & Electro-Optics*, pages E1–E7, 1998.

E. Atzeni and A. Salmi. Economics of additive manufacturing for end-usable metal parts. *The International Journal of Advanced Manufacturing Technology*, 62 (9–12): 1147–1155, 2012.

A. H. Azman, F. Vignat, and F. Villeneuve. Evaluating current CAD tools performances in the context of design for additive manufacturing. *Joint Conference on Mechanical, Design Engineering & Advanced Manufacturing*, 80: 1–7, 2014.

M. Bähr, J. Buhl, G. Radow, J. Schmidt, M. Bambach, M. Breuß, and A. Fügenschuh. Stable honeycomb structures and temperature based trajectory optimization for wire-arc additive manufacturing. *Optimization and Engineering*, pages 1–62, 2020.

L. Bai, J. Zhang, X. Chen, C. Yi, R. Chen, and Z. Zhang. Configuration optimization design of Ti6Al4V lattice structure formed by SLM. *Materials*, 11 (10): 1856, 2018.

E. Balas. An additive algorithm for solving linear programs with zero-one variables. *Operations Research*, 13 (4): 517–546, 1965.

A. Bandyopadhyay and B. Heer. Additive manufacturing of multi-material structures. *Materials Science and Engineering: Reports*, 129: 1–16, 2018.

M. W. Barclift and C. B. Williams. Examining variability in the mechanical properties of parts manufactured via polyjet direct 3D printing. In *International Solid Freeform Fabrication Symposium*, pages 6–8. University of Texas at Austin, 2012.

O. A. Bauchau and J. I. Craig. Euler-bernoulli beam theory. In *Structural Analysis*, pages 173–221, 2009.

J. Bauer, W. Gutkowski, and Z. Iwanow. A discrete method for lattice structures optimization. *Engineering Optimization*, 5 (2): 121–127, 1981.

M. S. Bazaraa, J. J. Jarvis, and H. D. Sherali. *Linear Programming and Network Flows*. John Wiley & Sons, 2011.

H. Becker. Entwicklung und Implementierung einer Softwareerweiterung zur CAD-Modellierung von additiv gefertigten Gitterstrukturen in Ansys SpaceClaim. In *Bachelor's thesis*. University of Siegen, 2020.

A. Ben-Tal and M. P. Bendsøe. A new method for optimal truss topology design. *SIAM Journal on Optimization*, 3 (2): 322–358, 1993.

A. Ben-Tal and A. Nemirovski. Potential reduction polynomial time method for truss topology design. *SIAM Journal on Optimization*, 4 (3): 596–612, 1994.

A. Ben-Tal and A. Nemirovski. Robust truss topology design via semidefinite programming. *SIAM Journal on Optimization*, 7 (4): 991–1016, 1997.

A. Ben-Tal and A. Nemirovski. Structural design. In *Handbook of Semidefinite Programming*, pages 443–467. Springer, 2000.

A. Ben-Tal, M. Kočvara, and J. Zowe. Two nonsmooth approaches to simultaneous geometry and topology design of trusses. In *Topology Design of Structures*, pages 31–42. Springer, 1993.

M. P. Bendsøe. *Topology Optimization*. Springer, 2009.

M. P. Bendsøe and N. Kikuchi. Topology and layout optimization of discrete and continuum structures. *AIAA Progress in Aeronautics and Astronautics Series*, 150, 1993.

M. P. Bendsøe and O. Sigmund. *Optimization of Structural Topology, Shape, and Material*, volume 414. Springer, 1995.

M. P. Bendsøe and O. Sigmund. *Topology Optimization: Theory, Methods, and Applications*. Springer Science & Business Media, 2013.

M. P. Bendsøe, A. Ben-Tal, and J. Zowe. Optimization methods for truss geometry and topology design. *Structural Optimization*, 7 (3): 141–159, 1994.

W. Bennage and A. Dhingra. Optimization of truss topology using tabu search. *International Journal for Numerical Methods in Engineering*, 38 (23): 4035–4052, 1995a.

W. Bennage and A. Dhingra. Single and multi-objective structural optimization in discrete-continuous variables using simulated annealing. *International Journal for Numerical Methods in Engineering*, 38 (16): 2753–2773, 1995b.

D. Bertsimas, D. B. Brown, and C. Caramanis. Theory and applications of robust optimization. *SIAM Review*, 53 (3): 464–501, 2011.

H. Bikas, P. Stavropoulos, and G. Chryssolouris. Additive manufacturing methods and modeling approaches: a critical review. *The International Journal of Advanced Manufacturing Technology*, 83 (1–4): 389–405, 2016.

J. Bland. Optimal structural design by ant colony optimization. *Engineering Optimization*, 33 (4): 425–443, 2001.

D. Bojczuk and A. Rębosz-Kurdek. Topology optimization of trusses using bars exchange method. *Bulletin of the Polish Academy of Sciences. Technical Sciences*, 60 (2): 185–189, 2012.

J. Bonet and R. D. Wood. *Nonlinear Continuum Mechanics for Finite Element Analysis*. Cambridge University Press, 1997.

L. Borchmann, P. Frohn-Sörensen, and B. Engel. In situ detection and control of wrinkle formation during rotary draw bending. *Procedia Manufacturing*, 50: 589–596, 2020.

D. Bourell, J. P. Kruth, M. Leu, G. Levy, D. Rosen, A. M. Beese, and A. Clare. Materials for additive manufacturing. *CIRP Annals*, 66 (2): 659–681, 2017.

E. Brandl, B. Baufeld, C. Leyens, and R. Gault. Additive manufactured Ti-6Al-4V using welding wire: comparison of laser and arc beam deposition and evaluation with respect to aerospace material specifications. *Phys. Procedia*, 5 (2): 595–606, 2010.

B. Brenken, E. Barocio, A. Favaloro, V. Kunc, and R. B. Pipes. Fused filament fabrication of fiber-reinforced polymers: A review. *Additive Manufacturing*, 21: 1–16, 2018.

M. Burns. *Automated Fabrication: Improving Productivity in Manufacturing*. Prentice Hall, 1993.

S. A. Burns. *Recent Advances in Optimal Structural Design*. American Society of Civil Engineers (ASCE), 2002.

C. V. Camp. Design of space trusses using big bang–big crunch optimization. *Journal of Structural Engineering*, 133 (7): 999–1008, 2007.

C. V. Camp and B. J. Bichon. Design of space trusses using ant colony optimization. *Journal of Structural Engineering*, 130 (5): 741–751, 2004.

I. Campbell, O. Diegel, J. Kowen, and T. Wohlers. *Wohlers Report 2018: 3D Printing and Additive Manufacturing State of the Industry: Annual Worldwide Progress Report*. Wohlers Associates, 2018.

V. Canellidis, J. Giannatsis, and V. Dedoussis. Efficient parts nesting schemes for improving stereolithography utilization. *Computer-Aided Design*, 45 (5): 875–886, 2013.

J. Cao, E. Brinksmeier, M. Fu, R. X. Gao, B. Liang, M. Merklein, M. Schmidt, and J. Yanagimoto. Manufacturing of advanced smart tooling for metal forming. *CIRP Annals*, 68 (2): 605–628, 2019.

B. E. Carroll, T. A. Palmer, and A. M. Beese. Anisotropic tensile behavior of Ti-6Al-4V components fabricated with directed energy deposition additive manufacturing. *Acta Materialia*, 87: 309–320, 2015.

A. Cerveira, A. Agra, F. Bastos, and J. Gromicho. New branch and bound approaches for truss topology design with discrete areas. In *Proceedings of the American Conference on Applied Mathematics. Recent Advances in Applied Mathematics*, pages 228–233, 2009.

K.-H. Chang. *Design Theory and Methods using CAD/CAE: The Computer Aided Engineering Design Series*. Academic Press, 2014.

Y.-M. Chen and J.-J. Liu. Cost-effective design for injection molding. *Robotics and Computer-Integrated Manufacturing*, 15 (1): 1–21, 1999.

. G. Cheng and X. Guo. ε-relaxed approach in structural topology optimization. *Structural Optimization*, 13 (4): 258–266, 1997.

G. Cheng and Z. Jiang. Study on topology optimization with stress constraints. *Engineering Optimization*, 20 (2): 129–148, 1992.

P. W. Christensen and A. Klarbring. *An Introduction to Structural Optimization*. Springer Science & Business Media, 2008.

C. K. Chua and K. F. Leong. *3D Printing and Additive Manufacturing: Principles and Applications of Rapid Prototyping*. World Scientific Publishing Company, 2014.

C. K. Chua, K. F. Leong, and C. S. Lim. *Rapid Prototyping: Principles and Applications*. World Scientific Publishing Company, 2010.

A. Clausen. *Topology Optimization for Additive Manufacturing*. Technical University of Denmark, 2016.

M. Cloots, A. Spierings, and K. Wegener. Assessing new support minimizing strategies for the additive manufacturing technology SLM. In *24th International SFF Symposium – An Additive Manufacturing Conference, University of Texas at Austin*, pages 631–643, 2013.

R. C. Coates, M. Coutie, and F. K. Kong. *Structural Analysis*. CRC Press, 2019.

B. P. Conner, G. P. Manogharan, A. N. Martof, L. M. Rodomsky, C. M. Rodomsky, D. C. Jordan, and J. W. Limperos. Making sense of 3D printing: Creating a map of additive manufacturing products and services. *Additive Manufacturing*, 1: 64–76, 2014.

G. Cowper. The shear coefficient in Timoshenko's beam theory. *Journal of Applied Mechanics*, 33 (2): 335–340, 1966.

CPLEX. IBM ILOG CPLEX 9.0 Rerence Manual. https://www.ibm.com/de-de/products/ilog-cplex-optimization-studio, 2014. Accessed February 2021.

CPLEX. IBM ILOG CPLEX 12.1 Rerence Manual. https://www.ibm.com/de-de/products/ilog-cplex-optimization-studio, 2019. Accessed February 2021.

L. S. Da Silva, R. Simões, and H. Gervásio. *Design of Steel Structures: Eurocode 3: Design of Steel Structures, Part 1–1: General Rules and Rules for Buildings*. John Wiley & Sons, 2012.

G. B. Dantzig. *Linear Programming and Extensions*. Princeton University Press, 1963.

G. B. Dantzig, D. R. Fulkerson, and S. M. Johnson. On a linear-programming, combinatorial approach to the traveling-salesman problem. *Operations Research*, 7 (1): 58–66, 1959.

R. de Borst, M. A. Crisfield, J. J. Remmers, and C. V. Verhoosel. *Nonlinear finite element analysis of solids and structures*. John Wiley & Sons, 2012.

B. de Saint-Venant. Mémoire sur la torsion des prismes [torsion of prism]. *Mémoires des Savants étrangers*, 14: 233–560, 1855.

R. R. de Souza, L. F. F. Miguel, R. H. Lopez, L. F. F. Miguel, and A. J. Torii. A procedure for the size, shape and topology optimization of transmission line tower structures. *Engineering Structures*, 111:162–184, 2016.

K. Deb and S. Gulati. Design of truss-structures for minimum weight using genetic algorithms. *Finite Elements in Analysis and Design*, 37 (5): 447–465, 2001.

B. Descamps and R. F. Coelho. The nominal force method for truss geometry and topology optimization incorporating stability considerations. *International Journal of Solids and Structures*, 51 (13): 2390–2399, 2014.

P. Dewhurst and G. Boothroyd. Early cost estimating in product design. *Journal of Manufacturing Systems*, 7 (3): 183–191, 1988.

A. Dhingra and W. Bennage. Topological optimization of truss structures using simulated annealing. *Engineering Optimization*, 24 (4): 239–259, 1995.

R. Diestel. *Graph Theory*. Graduate Texts in Mathematics. Springer, 3 edition, 2006.

DIN ISO 128-34:2002-05. Technical drawings—general principles of presentation—part 34: Views on mechanical engineering drawings (ISO 128-34:2001); german version DIN ISO 128-34:2002-05. Standard, *Deutsches Institut für Normung e.V.*, Berlin, GER, May 2002.

M. W. Dobbs and L. P. Felton. Optimization of truss geometry. *Journal of the Structural Division*, 95 (10): 2105–2118, 1969.

B. Dörig, T. Ederer, P. F. Pelz, M. E. Pfetsch, and J. Wolf. Gearbox design via mixed-integer programming. In *Proceedings of the VII European Congress on Computational Methods in Applied Sciences and Engineering*, 2016.

W. Dorn. Automatic design of optimal structures. *Journal de Mecanique*, 3: 25–52, 1964.

E. Dursun. personal communication, 2019–2020.

C. L. Dym and I. H. Shames. *Solid Mechanics*. Springer, 1973.

T. Ederer, U. Lorenz, A. Martin, and J. Wolf. Quantified linear programs: a computational study. In *Algorithms—ESA 2011*, pages 203–214. Springer, 2011.

S. Ehara and Y. Kanno. Topology design of tensegrity structures via mixed integer programming. *International Journal of Solids and Structures*, 47 (5): 571–579, 2010.

K. Eisemann. Linear programming. *Quarterly of Applied Mathematics*, 13 (3): 209–232, 1955.

L. Euler. *The Rational Mechanics of Flexible or Elastic Bodies 1638–1788: Introduction to Vol. X and XI*. Springer Science & Business Media, 1960.

A. Evgrafov. Simultaneous optimization of topology and geometry of flow networks. *Structural and Multidisciplinary Optimization*, 32 (2): 99–109, 2006.

A. Faramarzi and M. Afshar. A novel hybrid cellular automata–linear programming approach for the optimal sizing of planar truss structures. *Civil Engineering and Environmental Systems*, 31 (3): 209–228, 2014.

A. M. Faustino, J. J. Júdice, I. M. Ribeiro, and A. S. Neves. An integer programming model for truss topology optimization. *Investigação Operacional*, 26 (1): 11–127, 2006.

M. Fenton, C. McNally, J. Byrne, E. Hemberg, J. McDermott, and M. O'Neill. Automatic innovative truss design using grammatical evolution. *Automation in Construction*, 39: 59–69, 2014.

V. C. Finotto, W. R. da Silva, M. Valášek, and P. Štemberk. Hybrid fuzzy-genetic system for optimising cabled-truss structures. *Advances in Engineering Software*, 62: 85–96, 2013.

C. A. Floudas and P. M. Pardalos. *Frontiers in Global Optimization*, volume 74. Springer Science & Business Media, 2013.

D. Frank and G. Fadel. Expert system-based selection of the preferred direction of build for rapid prototyping processes. *Journal of Intelligent Manufacturing*, 6 (5): 339–345, 1995.

W. E. Frazier. Metal additive manufacturing: a review. *Journal of Materials Engineering and Performance*, 23 (6): 1917–1928, 2014.

P. Frohn-Sörensen, M. Geueke, T. T. Belay, C. Kuhnhen, M. Manns, and B. Engel. 3d printed prototyping tools for flexible sheet metal drawing. 2020.

M. B. Fuchs. Fully-stressed trusses. In *Structures and Their Analysis*, pages 371–375. Springer, 2016.

R. B. Fuller. *Synergetics: Explorations in the Geometry of Thinking*. Estate of R. Buckminster Fuller, 1982.

Y.-C. Fung, P. Tong, and X. Chen. *Classical and computational solid mechanics*, volume 2. World Scientific Publishing Company, 2017.

G. Galilei. *Two New Sciences*.Dover, 1914.

T. Gally, C. M. Gehb, P. Kolvenbach, A. Kuttich, M. E. Pfetsch, and S. Ulbrich. Robust truss topology design with beam elements via mixed integer nonlinear semidefinite programming. *Applied Mechanics and Materials*, 807: 229–238, 2015.

T. Gally, A. Kuttich, M. E. Pfetsch, M. Schaeffner, and S. Ulbrich. Optimal placement of active bars for buckling control in truss structures under bar failures. *Uncertainty in Mechanical Engineering III*, 885: 119–130, 2018.

A. H. Gandomi and X.-S. Yang. Benchmark problems in structural optimization. In *Computational Optimization, Methods and Algorithms*, pages 259–281. Springer, 2011.

W. Gao, Y. Zhang, D. Ramanujan, K. Ramani, Y. Chen, C. B. Williams, C. C. Wang, Y. C. Shin, S. Zhang, and P. D. Zavattieri. The status, challenges, and future of additive manufacturing in engineering. *Computer-Aided Design*, 69: 65–89, 2015.

J. Gardan. Additive manufacturing technologies: state of the art and trends. *International Journal of Production Research*, 54 (10): 3118–3132, 2016.

P. Gay, D. Blanco, F. Pelayo, A. Noriega, and P. Fernández. Analysis of factors influencing the mechanical properties of flat PolyJet manufactured parts. *Procedia Engineering*, 132: 70–77, 2015.

A. Gebhardt. *Understanding Additive Manufacturing*. Elsevier, 2011.

C. Gengdong and J. Zheng. Numerical performance of two formulations of truss topology optimization. *Acta Mechanica Sinica*, 10 (4): 326–335, 1994.

I. Gibson, D. W. Rosen, and B. Stucker. The use of multiple materials in additive manufacturing. In *Additive Manufacturing Technologies*, pages 436–449. Springer, 2010.

I. Gibson, D. W. Rosen, B. Stucker, et al. *Additive Manufacturing Technologies*. Springer, 2014.

I. Gibson, D. Rosen, and B. Stucker. Directed energy deposition processes. In *Additive Manufacturing Technologies*, pages 245–268. Springer, 2015.

R. F. Gibson. *Principles of Composite Material Mechanics*. CRC press, 2016.

F. Glover and S. Zionts. A note on the additive algorithm of Balas. *Operations Research*, 13 (4): 546–549, 1965.

H. M. Gomes. Truss optimization with dynamic constraints using a particle swarm algorithm. *Expert Systems with Applications*, 38 (1): 957–968, 2011.

R. E. Gomory. An algorithm for the mixed integer problem. Technical report, RAND Corporation, 1960.

R. E. Gomory. An algorithm for integer solutions to linear programs. *Recent Advances in Mathematical Programming*, 64 (260–302): 14, 1963.

M. S. Gonçalves, R. H. Lopez, and L. F. F. Miguel. Search group algorithm: a new metaheuristic method for the optimization of truss structures. *Computers & Structures*, 153: 165–184, 2015.

X. Gong, T. Anderson, and K. Chou. Review on powder-based electron beam additive manufacturing technology. In *International Symposium on Flexible Automation*, volume 45110, pages 507–515. American Society of Mechanical Engineers, 2012.

J. E. Gordon. *Structures: or Why Things don't Fall Down*. Penguin, 1978.

H. Greenberg. *Integer Programming*. Academic Press, 1971.

M. D. Grigoriadis. Optimal programming of lot sizes, inventories and labor allocations–a comment. *Management Science*, 12 (7): 622–625, 1966.

D. Gross, W. Hauger, J. Schröder, W. A. Wall, and N. Rajapakse. *Engineering Mechanics 1: Statics*, volume 1. Springer Science & Business Media, 2012.

S. Groth, B. Engel, and K. Langhammer. Algorithm for the quantitative description of freeform bend tubes produced by the three-roll-push-bending process. *Production Engineering*, 12 (3–4): 517–524, 2018.

W. Grzesik. *Advanced Machining Processes of Metallic Materials: Theory, Modeling and Applications*. Elsevier, 2008.

GUROBI. Gurobi optimizer reference manual. http://www.gurobi.com, 2019. Accessed February 2021.

K. Habermehl. *Robust Optimization of Active Trusses via Mixed-Integer Semidefinite Programming*. Verlag Dr. Hut, 2013.

T. Hagishita and M. Ohsaki. Topology optimization of trusses by growing ground structure approach. *Structural and Multidisciplinary Optimization*, 37 (4): 377–393, 2009.

R. Hague, I. Campbell, and P. Dickens. Implications on design of rapid manufacturing. *Journal of Mechanical Engineering Science*, 217 (1): 25–30, 2003.

P. Hajela and C.-Y. Lin. Genetic search strategies in multicriterion optimal design. *Structural Optimization*, 4 (2): 99–107, 1992.

K. Hamilton. Planning, preparing and producing: Walking the tightrope between additive and subtractive manufacturing. *Metal Additive Manufacturing*, 2: 39–56, 2016.

S. R. Hansen and G. N. Vanderplaats. Approximation method for configuration optimization of trusses. *AIAA Journal*, 28 (1): 161–168, 1990.

M. Hartisch. personal communication, 2016–2020.

M. Hartisch. *Quantified Integer Programming with Polyhedral and Decision-Dependent Uncertainty*. PhD thesis, University of Siegen, Germany, 2020.

M. Hartisch and U. Lorenz. Mastering uncertainty: towards robust multistage optimization with decision dependent uncertainty. In *PRICAI 2019: Trends in Artificial Intelligence*, pages 446–458, 2019a.

M. Hartisch and U. Lorenz. Game tree search in a robust multistage optimization framework: Exploiting pruning mechanisms. *arXiv preprint arXiv : 1811.12146; To appear in: 16th International Conference on Advances in Computer Games, ACG 2019*, 2019b.

M. Hartisch and U. Lorenz. Robust multistage optimization with decision-dependent uncertainty. *Operations Research Proceedings 2019: Selected Papers of the Annual International Conference of the German Operations Research Society (GOR)*, pages 439–445, 2020.

M. Hartisch, T. Ederer, U. Lorenz, and J. Wolf. Quantified integer programs with polyhedral uncertainty set. In *Computers and Games—9th International Conference, CG 2016*, pages 156–166. Springer, 2016.

M. Hartisch, C. Reintjes, T. Marx and U. Lorenz. Robust topology optimization of truss-like space structures. In *Pelz P.F., Groche P. (eds) Uncertainty in Mechanical Engineering. ICUME 2021. Lecture Notes in Mechanical Engineering*, pages 296–306, 2021. https://www.springerprofessional.de/robust-topology-optimization-of-truss-like-space-structures/19200210.

O. Hasançebi. Adaptive evolution strategies in structural optimization: Enhancing their computational performance with applications to large-scale structures. *Computers & Structures*, 86 (1–2): 119–132, 2008.

O. Hasançebi and S. K. Azad. An exponential big bang–big crunch algorithm for discrete design optimization of steel frames. *Computers & Structures*, 110: 167–179, 2012.

O. Hasançebi and F. Erbatur. Layout optimization of trusses using improved GA methodologies. *Acta Mechanica*, 146 (1–2): 87–107, 2001.

W. S. Hemp. Studies in the theory of Michell structures. In *Applied Mechanics*, pages 621–628. Springer, 1966.

W. S. Hemp. *Optimum Structures*. Clarendon Press, 1973.

K. Hiramoto, H. Doki, and G. Obinata. Optimal sensor/actuator placement for active vibration control using explicit solution of algebraic Riccati equation. *Journal of Sound and Vibration*, 229 (5): penalty0 1057–1075, 2000.

K. D. Hjelmstad. *Fundamentals of Structural Mechanics*. Springer US, 2 edition, 2005.

G. A. Holzapfel. *Nonlinear Solid Mechancis: A Continuum Approach for Engineering*. John Wiley & Sons, 2000.

N. Hopkinson and P. Dicknes. Analysis of rapid manufacturing – using layer manufacturing processes for production. *Proceedings of the Institution of Mechanical Engineers, Part C: Journal of Mechanical Engineering Science*, 217 (1): 31–39, 2003.

N. Hopkinson, R. Hague, and P. Dickens. *Rapid Manufacturing: an Industrial Revolution for the Digital Age*. John Wiley & Sons, 2006.

R. Horst and P. M. Pardalos. *Handbook of Global Optimization*, volume 2. Springer Science & Business Media, 2013.

R. Horst and H. Tuy. *Global Optimization: Deterministic Approaches*. Springer Science & Business Media, 2013.

S. H. Huang, P. Liu, A. Mokasdar, and L. Hou. Additive manufacturing and its societal impact: a literature review. *The International Journal of Advanced Manufacturing Technology*, 67 (5–8): 1191–1203, 2013.

R. Huiskes and S. J. Hollister. From structure to process, from organ to cell: recent developments of FE-analysis in orthopaedic biomechanics. *Journal of Biomechanical Engineering*, 115 (4B): 520–527, 1993.

C. W. Hull. Apparatus for production of three-dimensional objects by stereolithography. *United States Patent, Appl., No. 638905, Filed*, 1984.

F. Irgens. *Continuum Mechanics*. Springer Science & Business Media, 2008.

ISO 10303-242:2020-04. Industrial automation systems and integration—product data representation and exchange—part 242: Application protocol: Managed model-based 3D engineering. Standard, *International Organization for Standardization*, Geneva, CH, Apr. 2020.

ISO/ASTM 52900. Additive manufacturing—general principles—terminology (ISO/ASTM 52900:2015); german version en ISO/ASTM 52900:2017. Standard, *International Organization for Standardization, ASTM International*, Geneva, CH, June 2017.

ISO/ASTM 52921. Standard terminology for additive manufacturing—coordinate systems and test methodologies (ISO/ASTM 52921:2013); german version en ISO/ASTM 52921:2016. Standard, *International Organization for Standardization, ASTM International*, Geneva, CH, Jan. 2013.

B. H. Jared, M. A. Aguilo, L. L. Beghini, B. L. Boyce, B. W. Clark, A. Cook, B. J. Kaehr, and J. Robbins. Additive manufacturing: Toward holistic design. *Scripta Materialia*, 135: 141–147, 2017.

J. Jiang, X. Xu, and J. Stringer. Support structures for additive manufacturing: a review. *Journal of Manufacturing and Materials Processing*, 2 (4): 64, 2018.

S. R. Johnston, D. W. Rosen, M. Reed, and H. V. Wang.Analysis of mesostructure unit cells comprised of octet-truss structures. In *Proceedings of the The Seventeenth Solid Freeform Fabrication Symposium*, pages 228–233, 2006.

Y. Kanno. Topology optimization of tensegrity structures under self-weight loads. *Journal of the Operations Research Society of Japan*, 55 (2): 125–145, 2012.

Y. Kanno. Exploring new tensegrity structures via mixed integer programming. *Structural and Multidisciplinary Optimization*, 48 (1): 95–114, 2013a.

Y. Kanno. Topology optimization of tensegrity structures under compliance constraint: a mixed integer linear programming approach. *Optimization and Engineering*, 14 (1): 61–96, 2013b.

Y. Kanno. Global optimization of trusses with constraints on number of different cross-sections: a mixed-integer second-order cone programming approach. *Computational Optimization and Applications*, 63 (1): 203–236, 2016.

Y. Kanno and X. Guo. A mixed integer programming for robust truss topology optimization with stress constraints. *International Journal for Numerical Methods in Engineering*, 83 (13): 1675–1699, 2010.

A. Kaveh and T. Bakhshpoori. Optimum design of space trusses using Cuckoo search algorithm with Levy flights. *Iranian Journal of Science and Technology Transactions of Civil Engineering*, 37 (1): 1–15, 2013.

A. Kaveh and S. Javadi. Shape and size optimization of trusses with multiple frequency constraints using harmony search and ray optimizer for enhancing the particle swarm optimization algorithm. *Acta Mechanica*, 225 (6): 1595–1605, 2014.

A. Kaveh and A. Nasrollahi. Performance-based seismic design of steel frames utilizing charged system search optimization. *Applied Soft Computing*, 22: 213–221, 2014.

A. Kaveh and H. Rahami. Analysis, design and optimization of structures using force method and genetic algorithm. *International Journal for Numerical Methods in Engineering*, 65 (10): 1570–1584, 2006.

A. Kaveh and S. Talatahari. Particle swarm optimizer, ant colony strategy and harmony search scheme hybridized for optimization of truss structures. *Computers & Structures*, 87 (5–6): 267–283, 2009a.

A. Kaveh and S. Talatahari. A particle swarm ant colony optimization for truss structures with discrete variables. *Journal of Constructional Steel Research*, 65 (8–9): 1558–1568, 2009b.

A. Kaveh and S. Talatahari. Size optimization of space trusses using big bang–big crunch algorithm. *Computers & Structures*, 87 (17–18): 1129–1140, 2009c.

A. Kaveh and S. Talatahari. A discrete big bang–big crunch algorithm for optimal design of skeletal structures. *Asian Journal of Civil Engineering*, 11 (1): 103–122, 2010a.

A. Kaveh and S. Talatahari. Optimal design of skeletal structures via the charged system search algorithm. *Structural and Multidisciplinary Optimization*, 41 (6): 893–911, 2010b.

A. Kaveh and A. Zolghadr. Topology optimization of trusses considering static and dynamic constraints using the CSS. *Applied Soft Computing*, 13 (5): 2727–2734, 2013.

A. Kaveh and A. Zolghadr. Democratic PSO for truss layout and size optimization with frequency constraints. *Computers & Structures*, 130: 10–21, 2014.

A. Kaveh, B. F. Azar, and S. Talatahari. Ant colony optimization for design of space trusses. *International Journal of Space Structures*, 23 (3): 167–181, 2008.

S. H. Khajavi, J. Partanen, and J. Holmström. Additive manufacturing in the spare parts supply chain. *Computers in Industry*, 65 (1): 50–63, 2014.

N. Khot. Optimization of structures with multiple frequency constraints. *Computers & Structures*, 20 (5): 869–876, 1985.

U. Kirsch. Optimal topologies of structures. *Applied Mechanics Reviews*, 42 (8): 223–239, 1989a.

U. Kirsch. Optimal topologies of truss structures. *Computer Methods in Applied Mechanics and Engineering*, 72 (1): 15–28, 1989b.

U. Kirsch. Efficient reanalysis for topological optimization. *Structural Optimization*, 6 (3): 143–150, 1993a.

U. Kirsch. *Structural Optimization: Fundamentals and Applications*. Springer, 1993b.

U. Kirsch and G. I. N. Rozvany. Design considerations in the optimization of structural topologies. *Optimization of Large Structural Systems*, 231: 121–138, 1993.

A. Klarbring, J. Petersson, B. Torstenfelt, and M. Karlsson. Topology optimization of flow networks. *Computer Methods in Applied Mechanics and Engineering*,192 (35–36): 3909–3932, 2003.

M. Kočvara. Truss topology design with integer variables made easy. *Optimization Online, Tech. Rep*, 2010.

Y. Kok, X. P. Tan, P. Wang, M. Nai, N. H. Loh, E. Liu, and S. B. Tor. Anisotropy and heterogeneity of microstructure and mechanical properties in metal additive manufacturing: A critical review. *Materials & Design*, 139: 565–586, 2018.

J. Kövecses. Dynamics of mechanical systems and the generalized free-body diagram—part I: General formulation. *Journal of Applied Mechanics*, 75 (6), 2008a.

J. Kövecses. Dynamics of mechanical systems and the generalized free-body diagram—part II: Imposition of constraints. *Journal of Applied Mechanics*, 75 (6), 2008b.

J. Kranz, D. Herzog, and C. Emmelmann. Design guidelines for laser additive manufacturing of lightweight structures in TiAl6V4. *Journal of Laser Applications*, 27 (1): S14001, 2015.

M. Kripka. Discrete optimization of trusses by simulated annealing. *Journal of the Brazilian Society of Mechanical Sciences and Engineering*, 26 (2): 170–173, 2004.

J.-P. Kruth, M.-C. Leu, and T. Nakagawa. Progress in additive manufacturing and rapid prototyping. *CIRP Annals-Manufacturing Technology*, 47 (2): 525–540, 1998.

C. Kuhnhen, J. Knoche, J. Reuter, S. S. H. Al-Maeeni, and B. Engel. Hybrid tool design for a bending machine. *To appear in: 14th Conference on Intelligent Computation in Manufacturing Engineering CIRP ICME*, 2020.

R. Kureta and Y. Kanno. A mixed integer programming approach to designing periodic frame structures with negative Poisson's ratio. *Optimization and Engineering*, 15 (3): 773–800, 2014.

P. M. Kurowski. Teaching finite element analysis for design engineers. In *Proceedings of the Canadian Engineering Education Association*, pages 224–230, 2006.

A. Kuttich. *Robust Topology Optimization and Optimal Feedback Controller Design for Linear Time-Invariant Systems via Nonlinear Semidefinite Programming*. Sierke Verlag, 2018.

T.-H. Kwok, H. Ye, Y. Chen, C. Zhou, and W. Xu. Mass customization: reuse of digital slicing for additive manufacturing. *Journal of Computing and Information Science in Engineering*, 17 (2): 0210091–0210097, 2017.

R. Lakes. Foam structures with a negative Poisson's ratio. *American Association for the Advancement of Science*, 235: 1038–1041, 1987.

M. Langelaar. Topology optimization of 3D self-supporting structures for additive manufacturing. *Additive Manufacturing*, 12: 60–70, 2016.

E. L. Lawler and D. E. Wood. Branch-and-bound methods: A survey. *Operations Research*, 14 (4): 699–719, 1966.

R. Leal, F. Barreiros, L. Alves, F. Romeiro, J. Vasco, M. Santos, and C. Marto. Additive manufacturing tooling for the automotive industry. *The International Journal of Advanced Manufacturing Technology*, 92 (5–8): 1671–1676, 2017.

K. S. Lee and Z. W. Geem. A new structural optimization method based on the harmony search algorithm. *Computers & Structures*, 82 (9–10): 781–798, 2004.

K. S. Lee, Z. W. Geem, S.-H. Lee, and K.-W. Bae. The harmony search heuristic algorithm for discrete structural optimization. *Engineering Optimization*, 37 (7): 663–684, 2005.

A. C. Lemonge and H. J. Barbosa. An adaptive penalty scheme for genetic algorithms in structural optimization. *International Journal for Numerical Methods in Engineering*, 59 (5): 703–736, 2004.

A. S. Lewis and M. L. Overton. Eigenvalue optimization. *Acta Numerica*, 5 (1): 149–190, 1996.

G. K. Lewis and E. Schlienger. Practical considerations and capabilities for laser assisted direct metal deposition. *Materials & Design*, 21 (4): 417–423, 2000.

J. A. Lewis. Direct ink writing of 3D functional materials. *Advanced Functional Materials*, 16 (17): 2193–2204, 2006.

L. Li, Z. Huang, F. Liu, and Q. Wu. A heuristic particle swarm optimizer for optimization of pin connected structures. *Computers & Structures*, 85 (7–8): 340–349, 2007.

L. Li, Z. Huang, and F. Liu. A heuristic particle swarm optimization method for truss structures with discrete variables. *Computers & Structures*, 87 (7–8): 435–443, 2009.

P. N. J. Lindecke, H. Blunk, J.-P. Wenzl, M. Möller, and C. Emmelmann. Optimization of support structures for the laser additive manufacturing of TiAl6V4 parts. *CIRP Conference on Photonic Technologies*, 74: 53–58, 2018.

W. Lingyun, Z. Mei, W. Guangming, and M. Guang. Truss optimization on shape and sizing with frequency constraints based on genetic algorithm. *Computational Mechanics*, 35 (5): 361–368, 2005.

R. Liu, Z. Wang, T. Sparks, F. Liou, and J. Newkirk. Aerospace applications of laser additive manufacturing. In *Laser Additive Manufacturing*, pages 351–371. Elsevier, 2017.

G. H. Loh, E. Pei, D. Harrison, and M. D. Monzon. An overview of functionally graded additive manufacturing. *Additive Manufacturing*, 23: 34–44, 2018.

U. Lorenz and J. Wolf. Solving multistage quantified linear optimization problems with the alpha–beta nested Benders decomposition. *EURO Journal on Computational Optimization*, 3 (4): 349–370, 2015.

G.-C. Luh and C.-Y. Lin. Optimal design of truss-structures using particle swarm optimization. *Computers & Structures*, 89 (23–24): 2221–2232, 2011.

M. R. Maheri and M. Narimani. An enhanced harmony search algorithm for optimum design of side sway steel frames. *Computers & Structures*, 136: 78–89, 2014.

H. A. Mang and G. Hofstetter. *Festigkeitslehre [Mechanics of Materials]*. Springer, 2013.

T. R. Marchesi, R. D. Lahuerta, E. C. Silva, M. S. Tsuzuki, T. C. Martins, A. Barari, and I. Wood. Topologically optimized diesel engine support manufactured with additive manufacturing. *IFAC-PapersOnLine*, 48 (3): 2333–2338, 2015.

S. Mars. *Mixed-Integer Semidefinite Programming with an Application to Truss Topology Design*. Verlag Dr. Hut, 2014.

J. E. Marsden and T. S. Ratiu. *Introduction to Mechanics and Symmetry: a Basic Exposition of Classical Mechanical Systems*, volume 17. Springer Science & Business Media, 2013.

J. McKeown. Growing optimal pin-jointed frames. *Structural Optimization*, 15 (2): 92–100, 1998.

K. Mela. Resolving issues with member buckling in truss topology optimization using a mixed variable approach. *Structural and Multidisciplinary Optimization*, 50 (6): 1037–1049, 2014.

P. Mercelis and J.-P. Kruth. Residual stresses in selective laser sintering and selective laser melting. *Rapid Prototyping Journal*, 12 (5): 254–265, 2006.

R. Mertens, S. Clijsters, K. Kempen, and J.-P. Kruth. Optimization of scan strategies in selective laser melting of aluminum parts with downfacing areas. *Journal of Manufacturing Science and Engineering*, 136 (6): 0110121–0110127, 2014.

A. G. M. Michell. The limits of economy of material in frame-structures. *The London, Edinburgh, and Dublin Philosophical Magazine and Journal of Science*, 8 (47): 589–597, 1904.

L. F. F. Miguel and L. F. F. Miguel. Shape and size optimization of truss structures considering dynamic constraints through modern metaheuristic algorithms. *Expert Systems with Applications*, 39 (10): 9458–9467, 2012.

L. F. F. Miguel, R. H. Lopez, and L. F. F. Miguel. Multimodal size, shape, and topology optimisation of truss structures using the firefly algorithm. *Advances in Engineering Software*, 56: 23–37, 2013.

A. Morsi, B. Geißler, and A. Martin. Mixed integer optimization of water supply networks. In *Mathematical Optimization of Water Networks*, pages 35–54. Springer, 2012.

A. Mortazavi and V. Toğan. Simultaneous size, shape, and topology optimization of truss structures using integrated particle swarm optimizer. *Structural and Multidisciplinary Optimization*, 54 (4): 715–736, 2016.

T. M. Müller, P. Leise, I.-S. Lorenz, L. C. Altherr, and P. F. Pelz. Optimization and validation of pumping system design and operation for water supply in high-rise buildings. *Optimization and Engineering*, pages 1–44, 2020.

J. M. Mulvey, R. J. Vanderbei, and S. A. Zenios. Robust optimization of large-scale systems. *Operations Research*, 43 (2): 264–281, 1995.

T. Nakamura and M. Ohsaki. A natural generator of optimum topology of plane trusses for specified fundamental frequency. *Computer Methods in Applied Mechanics and Engineering*, 94 (1): 113–129, 1992.

T. D. Ngo, A. Kashani, G. Imbalzano, K. T. Nguyen, and D. Hui. Additive manufacturing (3D printing): A review of materials, methods, applications and challenges. *Composites Part B: Engineering*, 143: 172–196, 2018.

D. S. Nguyen and F. Vignat. A method to generate lattice structure for additive manufacturing. *International Conference on Industrial Engineering and Engineering Management (IEEM)*, pages 966–970, 2016.

F. Ning, W. Cong, J. Qiu, J. Wei, and S. Wang. Additive manufacturing of carbon fiber reinforced thermoplastic composites using fused deposition modeling. *Composites Part B: Engineering*, 80: 369–378, 2015.

N. Noilublao and S. Bureerat. Simultaneous topology, shape and sizing optimisation of a three-dimensional slender truss tower using multi-objective evolutionary algorithms. *Computers & Structures*, 89 (23–24): 2531–2538, 2011.

J. M. Oberndorfer, W. Achtziger, and H. R. Hörnlein. Two approaches for truss topology optimization: a comparison for practical use. *Structural Optimization*, 11 (3–4): 137–144, 1996.

Y. Oh, C. Zhou, and S. Behdad. Part decomposition and assembly-based (re) design for additive manufacturing: A review. *Additive Manufacturing*, 22: 230–242, 2018.

M. Ohsaki. Genetic algorithm for topology optimization of trusses. *Computers & Structures*, 57 (2): 219–225, 1995.

M. Ohsaki. *Optimization of Finite Dimensional Structures*. CRC Press, 2016.

M. Ohsaki, K. Fujisawa, N. Katoh, and Y. Kanno. Semi-definite programming for topology optimization of trusses under multiple eigenvalue constraints. *Computer Methods in Applied Mechanics and Engineering*, 180 (1–2): 203–217, 1999.

W. Orchard-Hays. *Advanced linear-programming computing techniques*. McGraw-Hill, 1968.

R. H. Parry. *Mohr Circles, Stress Paths and Geotechnics*. CRC Press, 2004.

S. N. Patnaik, A. S. Gendy, L. Berke, and D. A. Hopkins. Modified fully utilized design (MFUD) method for stress and displacement constraints. *International Journal for Numerical Methods in Engineering*, 41 (7): 1171–1194, 1998.

P. Pedersen. On the optimal layout of multi-purpose trusses. *Computers & Structures*, 2 (5–6): 695–712, 1972.

P. Pedersen. Optimal joint positions for space trusses. *Journal of the Structural Division*, 99 (12): 2459–2476, 1973.

P. F. Pelz, U. Lorenz, T. Ederer, S. Lang, and G. Ludwig. Designing pump systems by discrete mathematical topology optimization: The artificial fluid systems designer. *International Rotating Equipment Conference*, pages 1–10, 2012.

D. Pham, S. Dimov, and R. Gault. Part orientation in stereolithography.*The International Journal of Advanced Manufacturing Technology*, 15 (9): 674–682, 1999.

D. T. Pham and R. S. Gault. A comparison of rapid prototyping technologies. *International Journal of Machine Tools and Manufacture*, 38 (10–11): 1257–1287, 1998.

V. Popovich, E. Borisov, A. Popovich, V. S. Sufiiarov, D. Masaylo, and L. Alzina. Functionally graded inconel 718 processed by additive manufacturing: Crystallographic texture, anisotropy of microstructure and mechanical properties. *Materials & Design*, 114: 441–449, 2017.

L. Portolés, O. Jordá, L. Jordá, A. Uriondo, M. Esperon-Miguez, and S. Perinpanayagam. A qualification procedure to manufacture and repair aerospace parts with electron beam melting. *Journal of Manufacturing Systems*, 41: 65–75, 2016.

P. Pöttgen, T. Ederer, L. Altherr, U. Lorenz, and P. F. Pelz. Examination and optimization of a heating circuit for energy-efficient buildings. *Energy Technology*, 4 (1): 136–144, 2016.

W. Prager. A note on discretized Michell structures. *Computer Methods in Applied Mechanics and Engineering*, 3 (3): 349–355, 1974.

W. Prager. Optimal layout of cantilever trusses. *Journal of Optimization Theory and Applications*, 23 (1): 111–117, 1977.

W. Prager. Nearly optimal design of trusses. *Computers & Structures*, 8 (3–4): 451–454, 1978.

W. R. Priedeman Jr and A. L. Brosch. Soluble material and process for three-dimensional modeling, 2004. US Patent 6,790,403.

M. Pyrz. Discrete optimization of geometrically nonlinear truss structures under stability constraints. *Structural Optimization*, 2 (2): 125–131, 1990.

Z. Quan, A. Wu, M. Keefe, X. Qin, J. Yu, J. Suhr, J.-H. Byun, B.-S. Kim, and T.-W. Chou. Additive manufacturing of multi-directional preforms for composites: opportunities and challenges. *Materials Today*, 18 (9): 503–512, 2015.

H. Rahami, A. Kaveh, and Y. Gholipour. Sizing, geometry and topology optimization of trusses via force method and genetic algorithm. *Engineering Structures*, 30 (9): 2360–2369, 2008.

S. Rajan. Sizing, shape, and topology design optimization of trusses using genetic algorithm. *Journal of Structural Engineering*, 121 (10): 1480–1487, 1995.

S. Rajeev and C. Krishnamoorthy. Discrete optimization of structures using genetic algorithms. *Journal of Structural Engineering*, 118 (5): 1233–1250, 1992.

S. Rajeev and C. Krishnamoorthy. Genetic algorithms-based methodologies for design optimization of trusses. *Journal of Structural Engineering*, 123 (3): 350–358, 1997.

S. S. Rao. *Engineering Optimization: Theory and Practice*. John Wiley & Sons, 2019.

J. N. Reddy. *Energy Principles and Variational Methods in Applied Mechanics*. John Wiley & Sons, 2017.

K. F. Reinschmidt. Discrete structural optimization. *Journal of the Structural Division*, 97 (1): 133–156, 1971.

C. Reintjes and U. Lorenz. Mixed integer optimization for truss topology design problems as a design tool for AM components. *International Conference on Simulation for Additive Manufacturing*, 2: 193–204, 2019.

C. Reintjes and U. Lorenz. Bridging mixed integer linear programming for truss topology optimization and additive manufacturing. *Optimization and Engineering*, pages 1–45, 2020.

C. Reintjes and U. Lorenz. Algorithm-driven optimization of support-free truss-like structures in early-stage design for additive manufacturing. *PAMM*, 20 (1): e202000204, 2021. URL https://onlinelibrary.wiley.com/doi/abs/10.1002/pamm.202000204.

C. Reintjes, M. Hartisch, and U. Lorenz. Lattice structure design with linear optimization for additive manufacturing as an initial design in the field of generative design. *Operations Research Proceedings 2017: Selected Papers of the Annual International Conference of the German Operations Research Society (GOR)*, pages 451–457, 2018.

C. Reintjes, M. Hartisch, and U. Lorenz. Design and optimization for additive manufacturing of cellular structures using linear optimization. *Operations Research Proceedings 2018: Selected Papers of the Annual International Conference of the German Operations Research Society (GOR)*, pages 371–377, 2019.

C. Reintjes, M. Hartisch, and U. Lorenz. Support-free lattice structures for extrusion-based additive manufacturing processes via mixed-integer programming. *Operations Research Proceedings 2019: Selected Papers of the Annual International Conference of the German Operations Research Society (GOR)*, pages 465–471, 2020.

C. Reintjes, J. Reuter, M. Hartisch, U. Lorenz and B. Engel. Towards CAD-based mathematical optimization for additive manufacturing—designing forming tools for tool-bound bending. In *Pelz P.F., Groche P. (eds) Uncertainty in Mechanical Engineering. ICUME 2021. Lecture Notes in Mechanical Engineering*, pages 12–22, 2021. https://www.springerprofessional. de/towards-cad-based-mathematical-optimization-for-additive-manufac/19200200.

J. Reuter. personal communication, 2020.

U. T. Ringertz. On topology optimization of trusses. *Engineering Optimization*, 9 (3): 209–218, 1985.

U. T. Ringertz. A branch and bound algorithm for topology optimization of truss structures. *Engineering Optimization*, 10 (2): 111–124, 1986.

D. W. Rosen. Design for additive manufacturing: a method to explore unexplored regions of the design space. *Annual Solid Freeform Fabrication Symposium*, 18: 402–415, 2007.

D. W. Rosen. Computer-aided design for additive manufacturing of cellular structures. *Computer-Aided Design and Applications*, 4 (5): 585–594, 2013.

D. Roylance. *Mechanical properties of materials*. Massachusetts Institute of Technology, 2008.

G. I. N. Rozvany. Structural layout theory—the present state of knowledge. *Wiley Series in Numerical Methods in Engineering*, pages 167–195, 1984.

G. I. N. Rozvany. Difficulties in truss topology optimization with stress, local buckling and system stability constraints. *Structural Optimization*, 11 (3-4): 213–217, 1996a.

G. I. N. Rozvany. Some shortcomings in Michell's truss theory. *Structural Optimization*, 12 (4): 244–250, 1996b.

G. I. N. Rozvany. A critical review of established methods of structural topology optimization. *Structural and Multidisciplinary Optimization*, 37 (3): 217–237, 2009.

G. I. N. Rozvany. *Structural Design via Optimality Criteria: the Prager Approach to Structural Optimization*, volume 8. Springer Science & Business Media, 2012.

G. I. N. Rozvany. *Topology Optimization in Structural Mechanics*, volume 374. Springer, 2014.

G. I. N. Rozvany and W. Gollub. Michell layouts for various combinations of line supports—I. *International Journal of Mechanical Sciences*, 32 (12): 1021–1043, 1990.

J. Rychlewski. On Hooke's law. *Journal of Applied Mathematics and Mechanics*, 48 (3): 303–314, 1984.

Y. Saadlaoui, J.-L. Milan, J.-M. Rossi, and P. Chabrand. Topology optimization and additive manufacturing: Comparison of conception methods using industrial codes. *Journal of Manufacturing Systems*, 43: 178–186, 2017.

A. Saboori, D. Gallo, S. Biamino, P. Fino, and M. Lombardi. An overview of additive manufacturing of titanium components by directed energy deposition: microstructure and mechanical properties. *Applied Sciences*, 7 (9): 883, 2017.

E. Sachs, M. Cima, and J. Cornie. Three-dimensional printing: rapid tooling and prototypes directly from a CAD model. *CIRP annals*, 39 (1): 201–204, 1990.

E. Salajegheh and G. N. Vanderplaats. Optimum design of trusses with discrete sizing and shape variables. *Structural Optimization*, 6 (2): 79–85, 1993.

C. Schäfer. *Optimization Approaches for Actuator and Sensor Placement and Its Application to Model Predictive Control of Dynamical Systems*. Verlag Dr. Hut, 2015.

C. Schänzle, L. C. Altherr, T. Ederer, U. Lorenz, and P. F. Pelz. As good as it can be—ventilation system design by a combined scaling and discrete optimization method. *Proceedings of the FAN*, 2015.

J. Schelbert. *Discrete Approaches for Optimal Routing of High Pressure Pipes*. PhD thesis, Friedrich-Alexander-Universität Erlangen-Nürnberg, Germany, 2015.

H. M. Schey. *Div, Grad, Curl, and All That: an Informal Text on Vector Calculus*. WW Norton & Company, 2005.

A. Schrijver. *Theory of Linear and Integer Programming*. Wiley, 1998.

R. Sedaghati. Benchmark case studies in structural design optimization using the force method. *International Journal of Solids and Structures*, 42 (21–22): 5848–5871, 2005.

A. P. Seyranian and A. A. Mailybaev. *Multiparameter Stability Theory with Mechanical Applications*, volume 13. World Scientific Publishing Company, 2003.

O. Sigmund. *Design of Material Structures Using Topology Optimization*. PhD thesis, Technical University of Denmark, Denmark, 1994.

O. Sigmund, N. Aage, and E. Andreassen. On the (non-) optimality of Michell structures. *Structural and Multidisciplinary Optimization*, 54 (2): 361–373, 2016.

W. S. Slaughter. *The Linearized Theory of Elasticity*. Springer Science & Business Media, 2012.

M. Sonmez. Discrete optimum design of truss structures using artificial bee colony algorithm. *Structural and Multidisciplinary Optimization*, 43 (1): 85–97, 2011.

P. F. Sörensen, B. Mašek, M. F.-X. Wagner, K. Rubešová, O. Khalaj, and B. Engel. Flexible manufacturing chain with integrated incremental bending and Q-P heat treatment for on-demand production of AHSS safety parts. *Journal of Materials Processing Technology*, 275: 116312, 2020.

SpaceClaim Corporation. *SpaceClaim Add-in Style Guide*. Available at a charge, 2014.

SpaceClaim Corporation. *SpaceClaim Developers Guide*. Available at a charge, 2019.

M. Stolpe. On the reformulation of topology optimization problems as linear or convex quadratic mixed 0–1 programs. *Optimization and Engineering*, 8 (2): 163–192, 2007.

M. Stolpe. On some fundamental properties of structural topology optimization problems. *Structural and Multidisciplinary Optimization*, 41 (5): 661–670, 2010.

M. Stolpe. To bee or not to bee – comments on "discrete optimum design of truss structures using artificial bee colony algorithm". *Structural and Multidisciplinary Optimization*, 44 (5): 707–711, 2011.

M. Stolpe. Truss optimization with discrete design variables: a critical review. *Structural and Multidisciplinary Optimization*, 53 (2): 349–374, 2016.

M. Stolpe. Truss topology design by linear optimization. In *Advances and Trends in Optimization with Engineering Applications*, pages 13–25. SIAM, 2017.

M. Stolpe and A. Kawamoto. Design of planar articulated mechanisms using branch and bound. *Mathematical Programming*, 103 (2): 357–397, 2005.

K. Subramani. An analysis of quantified linear programs. In *Discrete Mathematics and Theoretical Computer Science*, pages 265–277. Springer Berlin Heidelberg, 2003.

K. Subramani. Analyzing selected quantified integer programs. In *International Joint Conference on Automated Reasoning*, pages 342–356. Springer, 2004.

L. Suhl and T. Mellouli. *Optimierungssysteme: Modelle, Verfahren, Software, Anwendungen [Optimization Systems: Models, Methods, Software, Applications]*. Springer-Verlag, 2009.

G. Sved and Z. Ginos. Structural optimization under multiple loading. *International Journal of Mechanical Sciences*, 10 (10): 803–805, 1968.

W. Tang, L. Tong, and Y. Gu. Improved genetic algorithm for design optimization of truss structures with sizing, shape and topology variables. *International Journal for Numerical Methods in Engineering*, 62 (13): 1737–1762, 2005.

M. Tawarmalani and N. V. Sahinidis. A polyhedral branch-and-cut approach to global optimization. *Mathematical Programming*, 103 (2): 225–249, 2005.

G. Tejani, V. Savsani, and S. Bureerat. *Truss Topology Optimization: A Review—Past, Present, and Future*. Scholars' Press, 2018a.

G. Tejani, V. Savsani, V. K. Patel, and P. V. Savsani. Size, shape, and topology optimization of planar and space trusses using mutation-based improved metaheuristics. *Journal of Computational Design and Engineering*, 5 (2): 198–214, 2018b.

D. Thomas. *The Development of Design Rules for Selective Laser Melting*. PhD thesis, University of Wales, United Kingdom, 2009.

D. S. Thomas and S. W. Gilbert. *Costs and cost effectiveness of additive manufacturing*. National Institute of Standards and Technology, 2014.

M. K. Thompson, G. Moroni, T. Vaneker, G. Fadel, R. I. Campbell, I. Gibson, A. Bernard, J. Schulz, P. Graf, B. Ahuja, et al. Design for additive manufacturing: Trends, opportunities, considerations, and constraints. *CIRP Annals*, 65 (2): 737–760, 2016.

S. Timoshenko. *History of Strength of Materials: with a Brief Account of the History of Theory of Elasticity and Theory of Structures*. Courier Corporation, 1983.

S. Timoshenko, J. M. Gere, and W. Prager. *Theory of Elastic Stability*. McGraw-Hill Book, 1962.

A. R. Toakley. Optimum design using available sections. *Journal of the Structural Division*, 94 (5): 1219–1244, 1968.

V. Toğan and A. T. Daloğlu. Optimization of 3D trusses with adaptive approach in genetic algorithms. *Engineering Structures*, 28 (7): 1019–1027, 2006.

B. Topping. Shape optimization of skeletal structures: a review. *Journal of Structural Engineering*, 109 (8): 1933–1951, 1983.

B. Topping. Mathematical programming techniques for shape optimization of skeletal structures. In *Shape and Layout Optimization of Structural Systems and Optimality Criteria Methods*, pages 349–375. Springer, 1992.

B. Topping. Topology design of discrete structures. In *Topology Design of Structures*, pages 517–534. Springer, 1993.

R. A. Toupin. Saint-Venant's principle. *Archive for Rational Mechanics and Analysis*, 18 (2): 83–96, 1965.

A. Tyas, M. Gilbert, and T. Pritchard. Practical plastic layout optimization of trusses incorporating stability considerations. *Computers & Structures*, 84 (3–4): 115–126, 2006.

R. Vaidya and S. Anand. Optimum support structure generation for additive manufacturing using unit cell structures and support removal constraint. *Procedia Manufacturing*, 5: 1043–1059, 2016.

E. van de Ven, R. Maas, C. Ayas, M. Langelaar, and F. van Keulen. Continuous front propagation-based overhang control for topology optimization with additive manufacturing. *Structural and Multidisciplinary Optimization*, 57 (5): 2075–2091, 2018.

L. Vandenberghe and S. Boyd. Semidefinite programming. *SIAM Review*, 38 (1): 49–95, 1996.

J. Vanek, J. A. G. Galicia, and B. Benes. Clever support: Efficient support structure generation for digital fabrication. In *Computer Graphics Forum*, volume 33, pages 117–125. Wiley Online Library, 2014.

VDI 3405-3-3. VDI Richtlinien, VDI 3405-3-3: Additive manufacturing processes, rapid manufacturing—design rules for part production using laser sintering and laser beam melting. Standard, *VDI-Gesellschaft Produktion und Logistik*, Düsseldorf, GER, Dec. 2015.

VDI 3405-3-4. VDI Richtlinien, VDI 3405-3-4: Additive manufacturing processes – design rules for part production using material extrusion processes. Standard, *VDI—Gesellschaft Produktion und Logistik*, Düsseldorf, GER, July 2019.

D. Wang, W. Zhang, and J. Jiang. Truss optimization on shape and sizing with frequency constraints. *AIAA journal*, 42 (3): 622–630, 2004.

D. Wang, Y. Yang, Z. Yi, and X. Su. Research on the fabricating quality optimization of the overhanging surface in SLM process. *The International Journal of Advanced Manufacturing Technology*, 65 (9–12): 1471–1484, 2013.

J. B. Weber and U. Lorenz. Optimizing booster stations. In *Proceedings of the Genetic and Evolutionary Computation Conference Companion*, pages 1303–1310, New York, NY, USA, 2017.

J. B. Weber, M. Hartisch, A. D. Herbst, and U. Lorenz. Towards an algorithmic synthesis of thermofluid systems. *Optimization and Engineering*, pages 1–56, 2020.

L. Wei, T. Tang, X. Xie, and W. Shen. Truss optimization on shape and sizing with frequency constraints based on parallel genetic algorithm. *Structural and Multidisciplinary Optimization*, 43 (5): 665–682, 2011.

J. A. Weiss, B. N. Maker, and S. Govindjee. Finite element implementation of incompressible, transversely isotropic hyperelasticity. *Computer Methods in Applied Mechanics and Engineering*, 135 (1–2): 107–128, 1996.

P. J. Withers and H. Bhadeshia. Residual stress. part 1—measurement techniques. *Materials Science and Technology*, 17 (4): 355–365, 2001a.

P. J. Withers and H. Bhadeshia. Residual stress. part 2–nature and origins. *Materials Science and Technology*, 17 (4): 366–375, 2001b.

J. Wolf. *Quantified Linear Programming*. PhD thesis, Technical University of Darmstadt, Germany, 2015.

L. A. Wolsey and G. L. Nemhauser. *Integer and Combinatorial Optimization*, volume 55. John Wiley & Sons, 1999.

P. Wriggers. *Nonlinear Finite Element Methods*. Springer Science & Business Media, 2008.

B. Xu, J. Jiang, W. Tong, and K. Wu. Topology group concept for truss topology optimization with frequency constraints. *Journal of Sound and Vibration*, 261 (5): 911–925, 2003.

Y. Yan and L. Yam. A synthetic analysis on design of optimum control for an optimized intelligent structure. *Journal of Sound and Vibration*, 249 (4): 775–784, 2002.

X.-S. Yang. Firefly algorithms for multimodal optimization. In *International Symposium on Stochastic Algorithms*, pages 169–178. Springer, 2009.

İ. Yanıkoğlu, B. L. Gorissen, and D. den Hertog. A survey of adjustable robust optimization. *European Journal of Operational Research*, 277 (3): 799–813, 2019.

E. Yasa, O. Poyraz, E. U. Solakoglu, G. Akbulut, and S. Oren. A study on the stair stepping effect in direct metal laser sintering of a nickel-based superalloy. *Procedia CIRP*, 45: 175–178, 2016.

D. Yates, A. Templeman, and T. Boffey. The complexity of procedures for determining minimum weight trusses with discrete member sizes. *International Journal of Solids and Structures*, 18 (6): 487–495, 1982.

K. Yonekura and Y. Kanno. Global optimization of robust truss topology via mixed integer semidefinite programming. *Optimization and Engineering*, 11 (3): 355–379, 2010.

Y. Zhang and K. Chou. A parametric study of part distortions in fused deposition modeling using three-dimensional finite element analysis. *Proceedings of the Institution of Mechanical Engineers, Part B: Journal of Engineering Manufacture*, 222 (8): 959–968, 2008.

M. Zhou. Difficulties in truss topology optimization with stress and local buckling constraints. *Structural Optimization*, 11 (2): 134–136, 1996.

Z. Zhu, V. G. Dhokia, A. Nassehi, and S. T. Newman. A review of hybrid manufacturing processes–state of the art and future perspectives. *International Journal of Computer Integrated Manufacturing*, 26 (7): 596–615, 2013.

Printed in the United States
by Baker & Taylor Publisher Services